# Fundamentals of Sustainable Aviation

*Fundamentals of Sustainable Aviation* is the first textbook to survey the critical field of sustainability within the aviation industry. Taking a systems thinking approach, it presents the foundational principles of sustainability and methodically applies them to different aviation sectors.

Opening with the basics of sustainability, emphasising the Sustainable Development Goals, the book then considers the environmental, economic and social dimensions of aviation. The following chapters apply these insights to aviation design, supply chains, operations, maintenance and facilities. The final chapter examines the concept of resilience in sustainable aviation. Overall, the textbook shows how future sustainability can be achieved by making better decisions today.

Students are supported with international case studies throughout the book. Slides, test questions and a teaching manual are available for instructors. This textbook is the ideal resource for courses on sustainable aviation globally and will also be of great interest to professionals in the field.

**Eva Maleviti** is Assistant Professor and Program Coordinator of the MS in sustainability in aviation and aerospace at Embry-Riddle Aeronautical University. She has been an academic for over 14 years in British, US and Greek Academic Institutions. Dr Maleviti has taught and developed various aviation-related courses in aviation sustainability, safety and quality, human factors in aviation maintenance, risk management, aviation legislation, environmental and energy management systems, and waste management practices for aerospace organisations. Since 2019, she has been a technical expert for the ICAO CORSIA and EU-ETS schemes with the Hellenic Accreditation System. For ten years, she was the academic coordinator and instructor at the Training Directorate of Hellenic Aerospace Industry (EASA Part 147 MTO). She has extensive experience as an aviation consultant in air operator and maintenance organisations, offering sustainability strategy development, ESG reporting and various tailor-made aviation consulting services. She is IRCA-certified Lead Auditor for Quality Management Systems ISO 9001 and in 9100D – Quality Management Systems for Aviation, Space and Defense Organizations. She holds a private pilot's license (A), a UAS license specific category C and is a member of the Royal Aeronautical Society, UK. She has a PhD and MSc in sustainable development, a graduate certificate in aviation maintenance and a BSc in physics.

## Aviation Fundamentals

Series Editor: Suzanne K. Kearns

*Aviation Fundamentals* is a series of air transport textbooks that incorporate instructional design principles to present content in a manner that is engaging to the learner, at an accessible level for young adults, allowing for practical application of the content to real-world problems via cases, reflection questions and examples. Each textbook will be supported by a companion website of supplementary materials and a test bank. The series is designed to help facilitate the recruitment and education of the next generation of aviation professionals (NGAP), a task which has been named a 'Global Priority' by the ICAO Assembly. It will also support education for new air transport sectors that are expected to rapidly evolve in future years, such as commercial space and the civil use of remotely piloted aircraft. The objective of *Aviation Fundamentals* is to become the leading source of textbooks for the variety of subject areas that make up aviation college/university degree programmes, evolving in parallel with these curricula.

**Fundamentals of International Aviation Law and Policy**
*Benjamyn I. Scott and Andrea Trimarchi*

**Fundamentals of Airline Operations**
*Gert Meijer*

**Fundamentals of International Aviation**
*Suzanne K. Kearns*

**Fundamentals of Airline Marketing**
*Scott Ambrose and Blaise Waguespack*

**Fundamentals of Statistics for Aviation Research**
*Michael A. Gallo, Brooke E. Wheeler and Isaac M. Silver*

**Fundamentals of Airport Planning**
Theory and Practice
*Lakshmanan Ravi*

**Fundamentals of Sustainable Aviation**
*Eva Maleviti*

For more information about this series, please visit: www.routledge.com/Aviation-Fundamentals/book-series/AVFUND

# Fundamentals of Sustainable Aviation

**Eva Maleviti**

LONDON AND NEW YORK

Designed cover image: Getty

First published 2024
by Routledge
4 Park Square, Milton Park, Abingdon, Oxon OX14 4RN

and by Routledge
605 Third Avenue, New York, NY 10158

*Routledge is an imprint of the Taylor & Francis Group, an informa business*



*British Library Cataloguing-in-Publication Data*
A catalogue record for this book is available from the British Library

ISBN: 978-1-032-16978-1 (hbk)
ISBN: 978-1-032-16407-6 (pbk)
ISBN: 978-1-003-25123-1 (ebk)

DOI: 10.4324/9781003251231

Typeset in Times New Roman
by Apex CoVantage, LLC

Access the Support Material: www.routledge.com/ 9781032169781

**To my parents Niki and George, for their continuous support and encouragement to accomplish all my goals!**

# Contents

# Figures

# Tables

# 1 Fundamentals of Sustainability

### Chapter Outcomes

At the end of this chapter, you will be able to do the following:

- Define the concept of sustainable development.
- Discuss the principles of sustainability.
- Identify the three pillars of sustainability.
- Explain the mutual preservation of the three pillars of sustainability.
- Understand the necessity to bring sustainability to aviation.

### Introduction

The aviation industry is an extensive and complex system with multiple interactions inside and outside its boundaries. Sustainability could be part of all aviation subsystems, and it could bring not only change but also benefits. Sustainability is a concept that should be applied holistically to a system, an organisation, a business or even how we think. The word holistic comes from the Greek word 'holos', which means 'a total', 'a whole'. *Holism* is the concept that requires all systems and their properties to be viewed as a complete element and not just as a collection of parts.[1] The idea of sustainability made a new entry into our lives very recently, and it applies, addresses and concerns almost all human activities. Before exploring the benefits of sustainability in the aviation industry, let us first examine the concept of sustainability, its origin and its presentation through the 17 Sustainable Development Goals (SDGs). We will link the general sustainability concept with some leading aviation organisations, such as ICAO, IATA and ATAG. Before we start drawing the pathway to aviation sustainability, it is essential to deepen in the sustainability basics and align them with commercial aviation and its multi-operational disciplines.

### The Concept of Sustainability

Like any other human activity and industrial sector, a sustainable commercial aviation should be able to address all three pillars equally through a series of actions and initiatives. Sustainable aviation can aim for economic growth for the industry, respect the natural environment and protect the social element, including people working in the industry and the local and global community where the industry operates. Everything starts with the definition of sustainable development. *Sustainable development* is development that meets the needs of the present without compromising the ability of future generations to meet their own needs.[2] In other words, sustainability is the practical aspect of actions to bring sustainable development to a system or entity.[3] The terms sustainability and sustainable development are often used interchangeably; even though the two concepts have a

DOI: 10.4324/9781003251231-1

common ground of origin, the notion of the two differs. However, 'sustainability' is the most used term, which we are using in this textbook.

> The concept of sustainable development was defined in the World Commission on Environment and Development's (WCED) 1987 Brundtland Report 'Our Common Future'.[2] The Brundtland Commission aimed to help world nations towards sustainable development. Then sustainable development became an essential concept in the vocabulary of politicians, practitioners and planners.

Before we move on to the analysis and connection of sustainability with aviation, we must build the foundation that will support us by going more profound to concepts and meanings relevant and related to the term. The report 'Our Common Future' is the cornerstone of sustainable development and sustainability. It is a document that first introduced the concept of sustainable development, prioritising the future and the next generations. The report shows that it is necessary to consider strategies supporting sustainable development, but it did not stay there. New government policies and support are necessary to protect the environment and humans and sustain economic growth. However, economic growth must be balanced with society and the environment.[2] The concept of sustainability showed the need for a significant change which will concern, embrace and benefit everyone. Not only governmental policies but actions that will benefit societies in need will support a balanced economic growth that will bring profits and assist measures that will protect the natural environment.

> Environmental science is not the same as sustainability. Environmental science has a focus on protecting the natural ecosystem. Sustainability supports preserving the natural ecosystem in a balanced approach with the economy and society.

***The Three Pillars of Sustainability***

The three pillars of sustainability are the environment, society and economy. These three elements are necessary to support sustainability. The term 'pillar' is used to show each one's necessity for supporting sustainability.

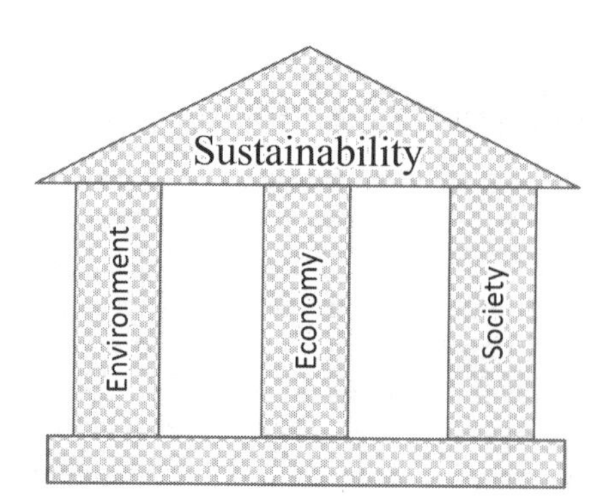


*Figure 1.1* The three pillars of sustainability

The environmental pillar refers to the protection of the environment and the natural ecosystem. The social pillar addresses a healthy society where humanity is at the centre. The economic pillar discourses economic growth, but this growth should maintain the other two pillars equally. However, because all three components are interlinked and interrelated, that is why they are represented as three intersecting cycles. All three of them connect and create subsections requiring a more profound analysis of sustainability. Remember that sustainability is a holistic approach, and the only way to succeed is to look closely at the interconnections between the different sustainability circles.

*First Circle – Environmental Sustainability*

The purpose of sensible use of natural resources, handling the environment and waste so the impacts from human activities are reduced to a minimum and promoting renewable energy sources are some elements included in the environmental circle of sustainability.

*Figure 1.2* The environment as element of sustainability

*Second Circle – Social Sustainability*

The social pillar concerns the impact of any activity from any entity on society. If the entity is a company, then the social pillar of sustainability examines the company's impact on society internally and externally. The internal society of a company has to do with the employees and its stakeholders. Healthy and fair working conditions, employee equity and equality, gender and race justice and work ethics are some components of social sustainability. The external society is the local society where the company is found, its customers and all external stakeholders involved in the company's activities.

*Figure 1.3* Social sustainability means people

*Third Circle – Economic Sustainability*

Any company, entity or society cannot be viable and long-lasting without economic viability. This means that profit and economic growth are necessary for survival. However, if economic growth does not support the other two pillars, the term economic sustainability is not valid. Throughout the years, we have seen cases where profit was dominant in business goals, neglecting the other two aspects, environment and society. These entities or their actions cannot be characterised as sustainable. Sustainability needs all three circles to be equally supported and represented at a business level. One cannot grow as a burden to the other. As mentioned, the interconnections between the three circles must also be examined closely to understand how to achieve sustainability at a deeper level.

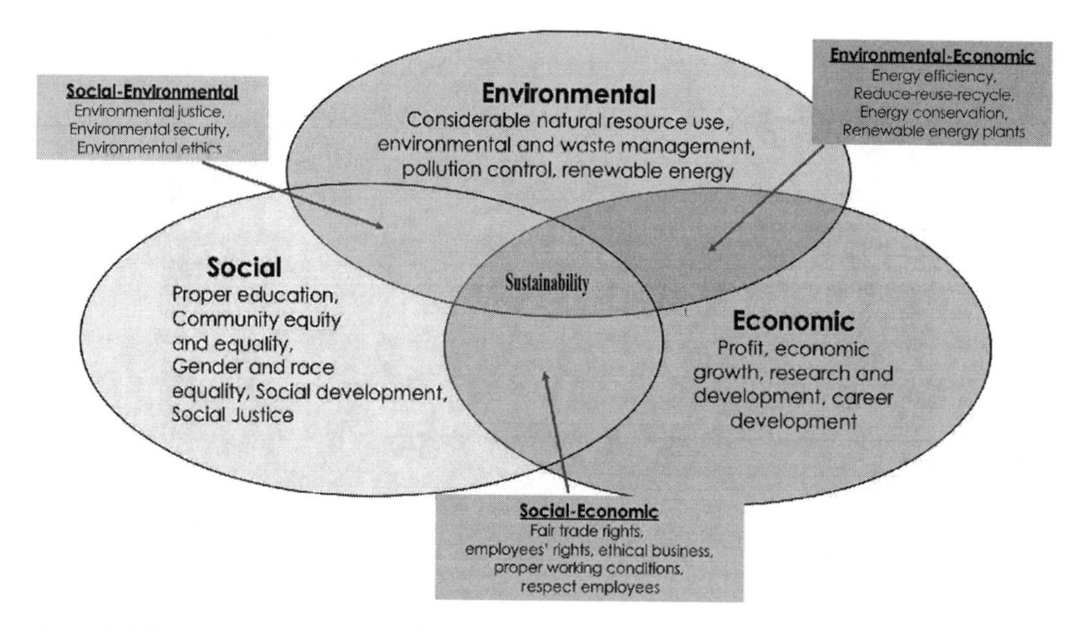


*Figure 1.4* The three circles of sustainability and their interrelations

The social-environmental connection includes environmental justice, ethics and security, among other components. This part shows the necessity of coexistence and equal consideration of activities that will not adversely affect society or the environment. Activities should consider minimising environmental effects and equally consider the growth and preservation of society. Similarly, environmental-economic activities include actions that will favour economic growth and respect and value the environment. Finally, the socioeconomic section considers economic and social growth as an outcome of business activities. As mentioned in the social sustainability component, the social component in a business environment must have a twofold approach: internal and external. The internal social component concerns employees and the way they are treated during their work. In an aviation environment, healthy conditions, appropriate resources and technical equipment according to the expected work outcomes, along with respect, inclusion, and development opportunities, are necessary elements for multiple and complex systems, such as aviation sector. The concept of the external social component refers to the effect of the company to the society and the local and global communities through its business. Corporate Social Responsibility is one element that is further described in the following chapters.

***The 17 Sustainable Development Goals***

Sustainable Development Goals (SDGs) have been first introduced in the 2030 Agenda for Sustainable Development. All the United Nations Member States adopted this agenda in 2015. This agenda aims to provide a shared plan among nations for peace, social justice and prosperity and protection of the planet. The core element of the Sustainable Development Goals calls for action between countries and building a global partnership. Ending hunger and poverty, improving health and education, reducing inequalities and supporting economic growth are some of the highlights of the 2030 Agenda. The United Nations provide significant support to meet the 17 SDGs.[4] Table 1.1

*Table 1.1* The 17 Sustainable Development Goals[4]

| *The 17 SDGs* | *Purpose* |
|---|---|
| No poverty | End poverty in all its forms everywhere |
| Zero hunger | End hunger, achieve food security and improved nutrition and promote sustainable agriculture |
| Good health and well-being | Ensure healthy lives and promote well-being for all at all ages |
| Quality education | Ensure inclusive and equitable quality education and promote lifelong learning opportunities |
| Gender equality | Achieve gender equality and empower all women and girls |
| Clean water and sanitation | Ensure availability and sustainable management of water and sanitation for all |
| Affordable and clean energy | Ensure access to affordable, reliable, sustainable and modern energy for all |
| Decent work and economic growth | Promote sustained, inclusive and sustainable economic growth, full and productive employment and decent work for all |
| Industry, innovation and infrastructure | Build resilient infrastructure, promote inclusive and sustainable industrialisation and foster innovation |
| Reduced inequalities | Reduce inequality within and among countries |
| Sustainable cities and communities | Make cities and human settlements inclusive, safe, resilient and sustainable |
| Responsible consumption and production | Ensure sustainable consumption and production patterns |
| Climate action | Take urgent action to combat climate change and its impacts |
| Life below water | Conserve and sustainably use the oceans, seas and marine resources for sustainable development |
| Life on land | Protect, restore and promote sustainable use of terrestrial ecosystems, sustainably manage forests, combat desertification and halt and reverse land degradation and biodiversity loss |
| Peace, justice and strong institutions | Promote peaceful and inclusive societies for sustainable development, provide access to justice for all and build effective, accountable and inclusive institutions at all levels |
| Partnerships for the goals | Strengthen the means of implementation and revitalise the global partnership for sustainable development |

lists the 17 SDGs with a short description of their purpose. The 17 SDGs show that sustainability should be part of all human activities, and all three pillars should be equally addressed. The achievement of these goals creates a world where inequalities do not exist; the economy can grow to support the people and provide at least the basic needs for survival and existence. Then the environment and the natural ecosystem are equally valued as essential components to respect and support through any human activity.

New technologies, infrastructure and industry are part of the goals scheme. Sustainability has become part of all business activities. The commercial aviation could not be left behind, and this need started quite recently to arise in the aviation sector. It is common knowledge that aviation is a very complex system with valuable economic and societal input. However, there is an extensive discussion about aviation's environmental effects. Aviation and aerospace science and engineering have lately focused on reducing environmental effects, especially on new technology that can reduce aircraft emissions. Although attention should not be given only to this aspect. As previously mentioned, sustainability should address all three circles. It is essential to explore the aviation sector closely and identify how sustainability should be applied, numbering the benefits that will arise from such a decision for the whole industry and its different operational domains.

## Sustainability in Aviation

Sustainability in aviation is a new term addressed recently and closely related to Sustainable Aviation Fuels (SAF) and sustainable designs in engines and airframes, designs with fewer emissions

*Figure 1.5* Airplane looking up

and aircraft types with reduced noise levels. However, sustainability in aviation is a lot wider concept. One aspect will concern the aircraft design and emissions and all the environmental actions and initiatives to reduce the effects of the sector on the natural environment. As mentioned in the previous section, the aviation business will be sustainable when all three circles are considered. Hence, addressing only the environmental aspect of sustainability does not provide the complete picture of how to make the aviation sustainable. Social sustainability should be addressed, considering employees, their working conditions and environment, the safe conditions of employees and customers, Corporate Social Responsibility and a positive return to society.

Similarly, economic sustainability must be addressed equally with the other two pillars. For every commercial business, economic growth is the most critical goal and factor of survival, but it should equally value the environment and people. If an aviation company aims only to profit without considering high safety standards for employees and customers or has no consideration for new schemes, initiatives, regulations and standards on how to reduce environmental effects, then this company is not sustainable, and it cannot become sustainable unless significant changes take place on their culture and how they contact their business. Each of the three sustainability pillars is closely analysed and examined in the following chapters to demonstrate how the aviation ecosystem can become a sustainable system. Before that, it is necessary to examine also what aviation regulatory authorities and organisations do and how they support the sustainability concept in aviation.

> Air Transport Action Group (ATAG) is a nonprofit association that focuses on applying long-term sustainability plans in commercial aviation. Its members are airports, airlines, aircraft design and production organisations and manufacturers, air traffic controller and pilot unions, air navigation companies and tourism and trade partners to name a few. One of ATAG's main activities is participating in the ICAO assemblies, working closely together for all aviation and environmental issues.[5] The role of ATAG is to bring together all sectors of the global aviation industry and provide a forum to address issues that affect us all. Lately, ATAG's work focuses on the aviation sustainability, climate change and SAF.[6] ATAG is an international association that, through its various activities, global research and industrial collaborations, supports the expansion of the air transport system, the creation of new work positions, trade, tourism and connectivity. It supports various remote communities with fast disaster response.[7] Needless to note that it is one of the most noticeable players that strongly supports sustainability in aviation.

> Air transport is at the heart of the global economic growth. It creates employment, facilitates trade, supports tourism and supports sustainable development across the world.

As a post–COVID-19 overview, the ATAG published their annual report pinpointing the five crucial steps in the industry as a recovery plan and as future goals to achieve.

a) **Review Traffic Forecasts**: The growth of aviation must be accompanied by coordinated actions that will limit environmental effects; governments must monitor this growth and try and mitigate any impacts. A shift to other means of transportation will significantly decrease the effects of the sector's growth. However, technology innovations and energy transition to low (zero) carbon emissions should be highly considered.

b) **Innovate with Technology**: Fuel efficiency is expected to improve by 20% to each generation of aircraft. Electric- and hydrogen-powered aircraft will serve regional, short-haul and/or medium-haul flights. Traditional fuels will still be available for aircraft that have not shifted to electric or hydrogen power and for long-haul flights but with a transition to 100% sustainable and low-carbon sources.
c) **Improve Operations and Infrastructure**: As part of the Waypoint Plan 2050, a wide range of measures to reduce CO2 emissions in airports, airlines and air traffic management are necessary. Operational efficiency is necessary, and continual improvement must be maintained.
d) **A Transition to SAF**: By 2050, aviation will need around 450 to 500 million tonnes of SAF yearly. This amount could be produced. However, sustainability criteria must be met, so food, land and water use are not compromised.
e) **Invest in Carbon Offsetting Measures**: Even though investment in out-of-sector carbon reduction is a shift aviation should pursue, there are other solutions to meet long-term goals. Offsets available in 2050 could be reduced since demand will also increase from other sectors.[8]

> Aviation and aerospace sustainability means to have an industry that aims to have a constant economic growth and profit along the years, respecting and valuing the social element, the people, by maintaining high safety standards, internally and externally, and at the same time protecting the natural environment that the industry is placed, aiming to operate with less environmental effects as possible and applying new efficient technologies.

***The Role of International Air Transport Association IATA***

The international air transport industry was born in 1919, with the signing of the Paris Convention. The same year, the International Air Transport Association (IATA) was established, the group that represents the majority of the world's airlines. The International Air Transport Association (IATA) has set various targets for contributing to a sustainable aviation sector. IATA has many suggestions and solutions to make the airline industry sustainable. IATA has an inclusive plan that embraces all aspects of sustainability through airline operations. Great emphasis is given again on environmental sustainability and SAF promotion; however, all the other two pillars are equally addressed. Evaluating the actions promoted by IATA, Figure 1.6 clusters these actions per each pillar of sustainability.

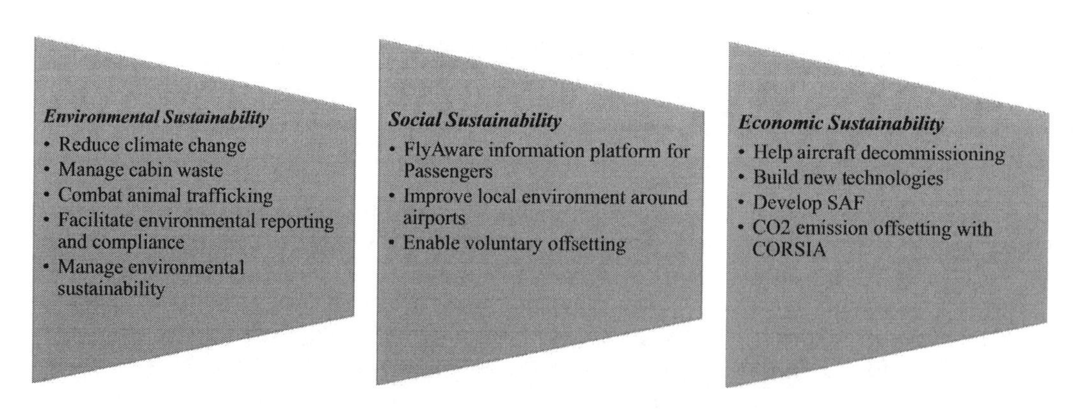


*Figure 1.6* IATA sustainability actions[9]

Looking beyond Figure 1.6, it is evident that actions cannot necessarily support only one pillar. Most of them support one or even the other two pillars of sustainability. To do a second evaluation of these actions and their benefits, they must be thoroughly evaluated and analysed. Each of the topics addressed in Figure 1.6 will look at them in detail in the following chapters.

### *International Civil Aviation Organization (ICAO) and Sustainability*

International Civil Aviation Organization (ICAO) plays a vital role in promoting and supporting sustainability in the aviation industry. ICAO is fully committed to promoting actions and schemes supporting the 17 Sustainable Development Goals. ICAO works closely with States and other United Nations bodies to show aviation's contribution to accomplish the 17 SDGs. ICAO has established strategic objectives that could cover and meet fifteen out of the 17 SDGs. Table 1.2 shows the alignment of the 17 SDGs and ICAO's strategic objectives. ICAO has developed a series of documents with long-term goals for creating an air transport framework for all Member States (MS) and supporting the development of an economically-viable civil aviation industry.

Table 1.2 shows that ICAO not only strongly supports sustainability in commercial aviation but it also provides a set of detailed actions and plans. These actions and plans address the society, the environment and the economic viability of the aviation and its Member States – areas with fewer privileges that are under numerous risks and other industries that rely on air transportation. It is also apparent that even though there is a great emphasis on SAF use to reduce aviation emissions, ICAO's policies equally address all three pillars for aviation sustainability, covering a wide range of the industry's operations.

## Conclusion

This chapter described sustainability and sustainable development and then provided a more detailed approach focusing on leading aviation organisations that promote sustainability. When someone opens the door to sustainability for the aviation, the focus is mainly on Sustainable Aviation Fuels and reducing aircraft emissions, a vital issue requiring attention, research and funding. However, sustainability for aviation could have a holistic perspective on addressing all three pillars: society, environment and economy. All three pillars must be equally addressed, measured and examined; otherwise, we cannot discuss a true sustainable system. Having set a solid foundation of sustainability fundamentals, the following chapters will address all three pillars in detail and expand their implementation to various aviation industry sectors but also what steps the aviation sector can take to swift to a true sustainable system. The main notion to follow is that a sustainable system, industry or company could seek ways to operate under the *holistic* approach, keeping in balance all three aspects; environment, society and economy.

## Key Points to Remember

- Sustainable development is the development that takes place now, without compromising the development of future generations.
- The three pillars of sustainability are the environment, the society and the economy.
- The UN developed the 17 Sustainable Development Goals, addressing all industrial sectors and requiring changes and adaptation from all.

*Table 1.2* ICAO and the UNSDGs

| *The 17 SDGs* | *ICAO Strategic Objectives* |
|---|---|
| No Poverty | ICAO facilitates States' investments in connectivity and infrastructure projects. These projects can support the expansion of tourism and trade. Provide economic support in less developed countries and emergency crises that require humanitarian aid. |
| Zero hunger | ICAO supports technological innovations to eliminate hunger across the globe. Unmanned aircraft systems and artificial intelligence are used to improve food quality in crop production and collect oil samples supporting agriculture precision. The Global Air Navigation Plan (GANP) follows a holistic approach towards aviation systems and supports increasing air navigation capacity and efficiency to accommodate rapid growth. Simultaneously, mitigate the risks associated with the operation of more aircraft within fixed airspace. The modernisation and harmonisation of the global air traffic management system improve the reliability of flight operations improving accessibility to remote areas and facilitating urgent access to sufficient food to all people, particularly those in vulnerable situations, constantly. Sustainable Aviation Fuels production could help to end hunger and all forms of malnutrition by enhancing the implementation of resilient agricultural practices. Also, some of the efforts that sustainable aviation actions can support are maintaining natural ecosystems and a more robust capacity of measures for climate change abatement, extreme weather, drought, flooding and other disasters. |
| Good health and well-being | ICAO and World Health Organization (WHO) have cooperated closely to eradicate and eliminate pandemics and diseases like Ebola and other communicable diseases. ICAO and WHO activities support strengthening the capacity of all Member States for early warning, risk reduction and management of national and global health risks. |
| Quality education | ICAO provides various training and events to support aviation professionals' development. ICAO leads the Next Generation of Aviation Professionals Programme (NGAP). It aims to ensure enough qualified and competent aviation professionals are available to operate, manage and maintain the future international air transport system. ICAO is in close cooperation with the United Nations Educational, Scientific and Cultural Organization (UNESCO), the International Labour Organization (ILO), the International Telecommunication Union (ITU) and UN Women in support of the NGAP Programme. ICAO and its stakeholders have established structures and mechanisms to expand the program. |
| Gender equality | The Gender Equality Programme (GEP) of ICAO promotes and enhances women's participation at all professional levels, managerial or technical, within the aviation sector. The GEP also supports States, regional, national and international organisations to support women's rights and equal representation in aviation. |
| Clean water and sanitation | N/A |
| Affordable and clean energy | As part of energy efficiency actions, ICAO's Global Air Navigation Plan (GANP) promotes actions for modern and improved global air traffic management. Efficient route design, airspace capacity and use of more efficient take-offs and landings using Performance-Based Navigation (PBN) bring advantages to minimise congestion and improve energy efficiency. Moreover, ICAO encourages Member States to invest in technologies and programs that can support the development of Sustainable Aviation Fuels and clean and renewable energy sources for the aviation sector. |
| Decent work and economic growth | ICAO supports the improvement of resource efficiency in all aviation production and operational activities while supporting economic growth without environmental degradation. The organisation supports growth as a part of aviation sustainability. In addition, aviation security can bring socioeconomic benefits to all ICAO Member States. The economic and financial cost to States and the aviation industry of terror attacks and security breaches can be significant, and the loss of human lives is unmeasurable. The transparency of the air transport regulatory framework is a crucial action supporting economic development, diversification, technological upgrade and innovation for everyone. |

| | |
|---|---|
| Industry, innovation and infrastructure | ICAO Global Aviation Safety Plan (GASP) promotes and supports civil aviation safety with frameworks that support regional, subregional and national plans. The GASP further supports the consistent implementation of safety among all ICAO Member States. Cybersecurity risks are also considered in ICAO's policies and provisions for international aviation operations. Unmanned Aircraft Systems (UAS) are significantly addressed within ICAO policies. In a sustainability aspect, the purpose is to assist those in need by delivering goods or supporting agricultural activities and the environment. |
| Reduced inequalities | The ICAO launched the 'No Country Left Behind' (NCLB) initiative as an opportunity for countries facing challenges implementing their Standards Recommended Practices (SARPs), plans, policies and programs. The NCLB focuses on assisting all Member States in developing based on equal opportunities. |
| Sustainable cities and communities | ICAO's Global Aviation Safety Plan supports the implementation of actions that strengthen national and regional development planning. ICAO develops guides on eco-friendly airports, evaluates policies for aircraft decommissioning and establishes Clean Development Mechanism (CDM) methodologies for aviation. These options will allow aviation projects to qualify for the generation of carbon credits under the CDM of the United Nations Framework Convention on Climate Change (UNFCCC). PBN, Airport Air Quality Manual, Environmental Technical Manual and the Eco-Airport e-collection support activities of sustainable urbanisation and human settlement planning at the national level, supporting at the same time the reduction of environmental impacts in cities, with improvements in air quality and waste management. |
| Responsible consumption and production | ICAO supports airport operations, eco-friendly activities and policies for aircraft recycling. It also promotes waste management actions that decrease environmental and health effects, air pollution and water and soil contamination. Strong partnerships with organisations such as the ILO, World Bank and UN World Trade Organization support similar actions that will reduce environmental effects and energy costs. |
| Climate action | ICAO provides internationally agreed policies, standards, guidance and tools for reducing or restricting the environmental impact of CO2 emissions from international aviation, including the development and implementation of a 'basket of measures' to meet the global goals of a 2% annual fuel efficiency improvement and carbon-neutral growth. A tool for control of emissions in airlines is the Carbon Offsetting Reduction Scheme for International Aviation (CORSIA). |
| Life below water | N/A |
| Life on land | Member States and the ICAO have been working closely with the industry to develop and deploy Sustainable Aviation Fuels (SAFs). SAF production must consider the land-use effects associated with their production. This effort must protect ecosystems and their sustainable use. For example, producing SAF from forestry remainings can be essential to sustainable forest management. |
| Peace, justice and strong institutions | ICAO helps and guides its Member States regarding standards and policies regarding global flight trafficking, conflict zones at risk and the use of small, unmanned aircraft, aiming to eliminate violence and decrease deaths worldwide. |
| Partnerships for the goals | ICAO assists its Member States in enhancing aviation safety, enabling the sustainable development of the air transport system resulting in economic growth. Safety improvements are part of the ICAO Universal Safety Oversight Audit Programme (USOAP). Additionally, ICAO has developed a strategic document for a possible Global Air Transport Plan (GATP), supporting partnerships among all the stakeholders for a sound and economically viable international civil aviation system. The primary goal is to contemporise the global air transport regulatory framework, with the long-term vision to promote tourism and trade among all ICAO Member States. |

- ICAO adapted the 17 Sustainable Development Goals, the fifteen Sustainable Development Goals concern the aviation industry.
- International organisations such as ICAO, IATA and ATAG and regulatory authorities such as EASA and FAA support the promotion of Sustainable Aviation Fuels.
- Sustainability in aviation concerns the reduction in environmental effects but also the economic and social sustainability in the whole aviation sector.
- The ATAG published their annual report pinpointing the five crucial steps in the industry as a recovery plan and as future goals to achieve.
- ICAO strongly supports sustainability in aviation and aerospace with detailed actions and plans.
- ICAO actions and plans include the society, the environment and the economic viability of the aviation industry and its Member States – areas with fewer privileges that are under numerous risks and other industries that rely on aviation activities

**Case Study: Pet Rescue Mission**[10, 11]

*Figure 1.7* Heart paw

In 2020, while COVID-19 was in spur, air transportation was in a long halt for months. The lack of air travel and all travel restrictions also terminated all flights, commercial or not. Under that circumstances, pet rescue missions were also halted, leading to overcrowding in numerous animal and pet rescue shelters. While on holidays abroad, people often find abandoned animals and adopt them. It has become a trend lately. Inevitably, as tourism has ceased, any pet rescue activity was also limited. This led many communities to handle a surplus of unadopted animals and overcrowded shelters, particularly in island states that rely heavily on air transport as a vital lifeline to the mainland. Therefore, COVID-19 restrictions also hindered cross-border pet rescue missions that would otherwise be responsible for relieving local shelters. This kind of missions depends on air travel and support from travellers who volunteer to transport animals while visiting different countries.

On the other hand, in specific locations, there was an increase in animal adoptions. People started working remotely, from home, for more than a year. Sadly, at the same time, regions saw an increase in abandoned pets due to the financial difficulties of their owners. Air transport bridges the gap between these two worlds, ensuring enough supply to meet the high demand in different regions. Without the significant activity of transportation, various animal rescue transfer programs have been suspended for months, which has prevented animals from being carried from island states. Pet rescue organisations have relied on local foster families to care for these animals without this option. The nonprofit organisation Paws Across the Pacific primarily transfers abandoned and rescued pets to their homes. In October 2020, Paws Across the Pacific hired a Hercules C-130 to fly to the Hawaiian Islands and carry 600 dogs and cats to their new homes. This flight was urgently needed to make space in Hawaii's shelters for at-risk pets who otherwise would not be able to receive the necessary care to survive. Of course, the airlift was done in coordination with the Hawaii Veterinary Medical Association to ensure they all arrived safely. When animals' living conditions are not as they should be, they can eventually affect human life too. When access to medical treatment is unavailable, as happened while COVID-19, animals and human life may be at threat. There is a thin line between how animal life can affect and threaten human life, especially in remote areas. It should not be neglected, though, that animals deserve to be taken care of and find a forever home. We are all part of the same environment, after all!

*Table 1.3* Acronym rundown

| | |
|---|---|
| ATAG | Air Transport Association Group |
| CDM | Clean Development Mechanisms |
| CORSIA | Carbon Offsetting Reduction Scheme for International Aviation |
| EASA | European Aviation Safety Agency |
| FAA | Federal Aviation Administration |
| MS | Member States |
| GANP | Global Air Navigation Plan |
| GEP | Gender Equality Programme |
| GASP | Global Aviation Safety Plan |
| GATP | Global Air Transport Plan |
| IATA | International Air Transport Association |
| ICAO | International Civil Aviation Organization |
| ILO | International Labor Organization |
| ITU | International Telecommunication Union |
| NGAP | Next Generation of Aviation Professionals Programme |
| NCLB | No Country Left Behind |
| PBN | Performance-Based Navigation |
| SAF | Sustainable Aviation Fuels |
| SARPs | Standards Recommended Practices |
| SDGs | Sustainable Development Goals |
| UAS | Unmanned Aircraft Systems |
| UN | United Nations |
| UNESCO | United Nations Educational, Scientific and Cultural Organization |
| USOAP | Universal Safety Oversight Audit Programme |
| WCED | World Commission on Environment and Development's |
| WHO | World Health Organization |

## Chapter Review Questions

1.1 How do you perceive the term 'sustainable aviation'? Give a brief and inclusive description.

1.2 What are the three pillars of sustainability, and what do they entail?

1.3 Why do you think that sustainability should apply to aviation industry and operations?

1.4 Address three Sustainable Development Goals and explain how they link to aviation.

1.5 Why is sustainability important for the aviation industry?

1.6 What benefits can aviation gain by being sustainable?

1.7 Research an example of an aviation sustainability case and explain how the three pillars of sustainability are met.

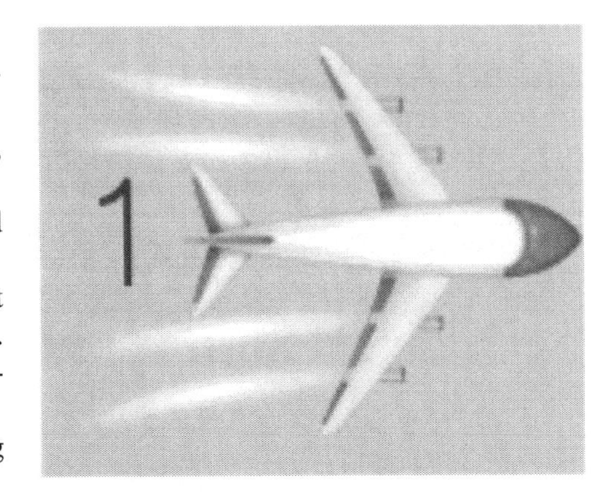


*Figure 1.8* Airplane looking right

1.8 Why can the review of traffic forecasts be a factor to support aviation sustainability?
1.9 Why is the Gender Equality Programme of ICAO necessary for aviation sustainability?
1.10 Group the 17 SDGs from Table 1.1 in each one of the three circles of sustainability of Figure 1.2.

***Case Study Questions***

1.11 In the pet rescue case study, a significant issue has been raised. When animals are left to the shelters under unsafe conditions, this is a situation that can compromise not only the pets but also humans. Explain how the safe conditions of pets while in shelters are part of environmental sustainability.
1.12 How does environmental sustainability connect with the animals, their protection and their rescue?
1.13 Why is a pet rescue mission from air transportation part of environmental sustainability?
1.14 What are the main factors that support sustainability not only the environmental but also the other two pillars in a pet rescue mission?
1.15 How does a pet rescue mission affect social sustainability?

## References


[1] Cambridge Dictionary. (n.d.). https://dictionary.cambridge.org/dictionary/english/holistic
[2] United Nations. (1987). *Report of the world commission on environment and development: Our common future*. United Nations- Sustainable Development. https://sustainabledevelopment.un.org/content/documents/5987our-common-future.pdf
[3] Maleviti, E., Atsarou, K., & Stamoulis, E. (2021, August 1–3). *Sustainability in aviation and aerospace design in the 9th global conference in global warming*. Online – Hosted by the University of Split
[4] United Nations. (2015). *The 17 goals | sustainable development*. The 17 Goals- Sustainable Development. https://sdgs.un.org/goals
[5] Air Transport Action Group. (2021). *Who we are*. Retrieved June 1, 2021, from www.atag.org/about-us/who-we-are.html
[6] ATAG. (2021a). *Activities overview*. ATAG – Activities Overview. www.atag.org/our-activities/activities-overview.html
[7] Air Transport Action Group. (2020, September 30). *Aviation: Benefits beyond borders 2020*. Aviation: Benefits Beyond Borders. https://aviationbenefits.org/downloads/aviation-benefits-beyond-borders-2020/
[8] Air Transport Action Group. (2020, September). *BluePrint for a green recovery*. https://aviationbenefits.org/media/167142/bgr20_final.pdf
[9] IATA. (n.d.). *Our actions for the environment*. IATA – Environment. Retrieved June 1, 2021, from www.iata.org/en/programs/environment/
[10] Air transport Action Group. (n.d.). *Rescuing pets across the pacific*. Retrieved February 18, 2022, from https://aviationbenefits.org/case-studies/rescuing-pets-across-the-pacific/
[11] Padgett, L. C. (2020, October 31). *More than 600 dogs and cats are airlifted out of Hawaii for new homes on the US mainland*. CNN. Retrieved February 18, 2022, from https://edition.cnn.com/2020/10/31/us/hawaii-paws-across-the-pacific-trnd/index.html

# 2 Environmental Sustainability in Aviation

### Chapter Outcomes

At the end of this chapter, you will be able to do the following:

- Explain the different types of aviation emissions.
- Assess the environmental impact of aviation.
- Learn about the EU-ETS and CORSIA emission schemes.
- Explain various environmental regulatory initiatives for aviation.
- Identify the environmental effects of the air transportation.

## Introduction

Aviation supports a vast array of business activities and offers a different quality of life, allowing people to visit friends and relatives, travel, experience new places, make distances smaller and make the world smaller. As discussed in Chapter 1, aviation supports tourism and economic activity, and at the same time, it supports remote areas in need or emergencies. A significant part of goods and merchandise are carried via air transport to several commercial hubs across the world's airports. Aviation must be sustainable, supporting society and the economy but, at the same time, operate harmoniously within the constraints imposed by the need for clean air and water, limit noise impacts and support a livable climate. Environmental sustainability requires the aviation industry to adopt practices that reduce environmental impacts and support preserving the natural environment. The Intergovernmental Panel on Climate Change (IPCC) is the international body responsible for assessing climate change. It was founded in 1988 by the United Nations Environmental Programme (UNEP) and the World Meteorological Organization (WMO). The purpose of this organisation is to advise and consult policymakers with information about climate change, identify potential climate risks and suggest options for adapting and mitigating environmental impacts. In 2018, the IPCC published a report on the effects of global warming and the temperature rise of 1.5°C above the preindustrial levels. As of the report, human activities are the main contributor to global warming and responsible for a 0.2°C per decade. As a primary mitigation measure and to control the temperature's increase, IPCC's suggestion is to reduce CO2 emissions from all human activities. The main goal is to mitigate CO2 emissions by 45% (2010 levels) by 2030, reaching net zero by 2050. As part of this agenda, aviation needs to proceed with specific measures and mitigation actions to reach that goal.[1, 2]

DOI: 10.4324/9781003251231-2

In 1997, the first legally binding requirements for greenhouse gas reduction were set by the Kyoto Protocol, addressing to all developed countries for greenhouse gas reduction and limitation. Emissions trading schemes were also firstly introduced in Kyoto Protocol. Initially, though, the aviation industry was excluded from these requirements and schemes, but at a later stage, ICAO was designated to regulate that matter with the establishment of the CORSIA Scheme.

**Aviation Environmental Impacts**

Some environmental impacts from aviation activities are air and noise pollution, water pollution and ground pollution, among others. Local communities might be exposed to noise from aircraft due to the vicinity. Streams, rivers and wetlands may be exposed to pollutants discharged in storm water runoff from airports or maintenance facilities, and aircraft engines emit pollutants into the air. In addition, emissions associated with commercial aviation may affect people's health and welfare, impacting the social element's quality of life, or in simple words, affecting the conditions of people living in neighbouring areas of aviation operations and activities. For example, if an area's economic activity relies on tourism and is highly impacted by airport noise or poor air quality, tourism activity and economic revenues will be reduced. Nevertheless, this is just one example of how aviation may impact the natural and social environment. The point is that poor environmental sustainability will affect both social and economic sustainability. To find the solution to that problem is first essential to define aviation emissions and the types of particulate matter discharged by direct aircraft operations. Emissions produced by aircraft engines are almost the same as emissions from fossil fuel internal combustion engines. A significant difference, however, is that a vast amount of aircraft emissions is produced at an altitude during the cruising stage of flight. These emissions, even at altitude, raise environmental concerns regarding their impact and effect on local air quality at ground level. Apart from emissions from main aircraft engines, there are emissions from the Auxiliary Power Unit (APU) while in the air. Also, while on the ground, emissions are released from the APU, the Ground Power Units (GPU) and Ground Support Equipment (GSE). Therefore, aviation environmental impacts derive from multiple sources, so when we aim to mitigate emissions or noise, we must be accurate in identifying the operations and the generating sources.

*Figure 2.1* Airplane positioning to take off

An Auxiliary Power Unit (APU) is a small stand-alone jet engine generally located in the aircraft's tail cone but, in some cases, is in an engine nacelle or the wheel well. The APU can start utilising only the aircraft battery(s) and, once running, will provide electrical power to aircraft systems and bleed air for air conditioning and engine start. When the APU is certified for use in flight, it can be used, as required, to provide an additional source of electrical power in the event of the complete loss of the main engine or its generator. Many airports, when near residential areas, include measures to limit the use of the APU due to noise[3]

A Ground Power Unit (GPU) is a fixed or mobile unit that connects to an aircraft's electrical system while on the ground. It provides either 120V AC or 28V DC power. Ground power units usually consist of a generator powered by a diesel engine.[4]

Ground Support Equipment (GSE) is the service and maintenance equipment used in airports to support operations and related activities while the aircraft is in the apron area. Baggage tugs, belt loaders, cargo loaders, forklifts, fuel trucks, lavatory trucks and pushback tractors are some of the types of vehicles that are part of this definition.[5]

***Aviation Emissions***

In 1999, ICAO, with the Intergovernmental Panel on Climate Change (IPCC) and in collaboration with the Scientific Assessment Panel to the Montreal Protocol on Substances that Deplete the Ozone Layer, requested the development of an assessment of aviation's contribution to international air pollution. Two fundamental points worth mentioning in this report. The first is the contribution of aircraft emissions and particles to climate change since they affect the atmospheric concentration of greenhouse gases, triggering condensed trails and increasing cirrus clouds. The second point is that aircraft emissions contribute to about 3.5% of the total radiative forcing by human activities, with the radiative forcing projected to grow.[6]

Radiative force is the difference between incoming and outgoing radiation. It is the most common indicator to quantify climate change. It is measured in Watts per square meter ($W/m^2$).[7, 8]

As mentioned earlier in this section and to be inclusive of aviation environmental impacts, we should focus on more than just the emissions associated directly with aircraft flights. In airports, apart from aircraft, cars, trucks and other vehicles are responsible for emissions due to fuel combustion, such as APU, GSE and GPU, as previously mentioned. Therefore, various emission streams and air-polluting substances are related to aviation operations. Table 2.1 briefly describes operations and aviation emissions with associated impacts.[9]

Aircraft engines generate carbon dioxide (CO2) and water vapor at 70% and 30%, respectively. Other pollutants such as nitrogen oxides, oxides of sulphur, carbon monoxide and hydrocarbons and particulate matter comprise less than 1%.[9]

Water vapor is one of the major players in climate change. Increase of water vapor increases temperature, causing more water to be absorbed into the air. Increase of temperature and water absorption increase are in a spiraling cycle.[10]

*Table 2.1* Aviation emissions

| *Emission Substances* | *Description* | *Emission Sources* | *Impacts* |
|---|---|---|---|
| **CO2** | CO2 is produced during the complete internal combustion of fossil fuels, such as HC, diesel, jet fuel and others. | - Aircraft<br>- APU<br>- GSE/GPU<br>- Airport Vehicles | - Global Warming<br>- Climate Change |
| **H2O** | Water vapour is a product from complete combustion of hydrogen in the fuels, with oxygen in the air. It produces the water vapour in the condensation trails or contrails. | - Aircraft<br>- APU<br>- GSE/GPU<br>- Airport Vehicles | - Global Warming<br>- Climate Change |
| **NOx** | NOx is formed during a combustion when nitrogen is combined with fuel. The nitrogen is oxidised and it creates NOx. | - Aircraft<br>- APU<br>- GSE/GPU<br>- Airport Vehicles | - Air Quality<br>- Climate Change |
| **HC** | HC are organic emissions formed by incomplete combustion of hydrocarbon fuel. Some of the HC emissions are toxic and hazardous. | - Aircraft<br>- APU<br>- GSE/GPU<br>- Airport Vehicles | - Air Quality<br>- Ozone Layer |
| **CH4** | CH4 has the most basic hydrocarbon form, and it is not a product of inflight emissions. | - APU<br>- GSE/GPU<br>- Airport Vehicles | - Air quality<br>- Climate Change |
| **CO** | CO is formed by an incomplete fuel carbon combustion of the carbon and is one of the most common factors to ozone formation. | - Aircraft<br>- APU<br>- GSE/GPU<br>- Airport Vehicles | - Air quality |
| **SOx** | SOx are generated when small quantities of sulphur in petroleum fuels, react with oxygen during combustion. | - Aircraft<br>- APU<br>- GSE<br>- Other airport equipment | - Air quality<br>- Climate Change |
| **Particulate matter** | Small particles form from black carbon that result from incomplete combustion and aerosols from condensed gases. | - Aircraft<br>- APU<br>- GSE/GPU<br>- Airport Vehicles | - Air quality<br>- Climate Change |

*Note*: Description of substances emitted from different aviation sources[7, 9]

Aircraft emissions and pollutants are formed differently near the earth's surface or higher. Specifically, 10% of emissions are generated close to the earth's surface, in height less than 3,000 feet above ground level, and 90% are emitted at altitudes higher than 3,000 feet. The height where CO and HCs are emitted differs; they are generated when aircraft engines operate at their lowest combustion efficiency, splitting the emissions into 30% below 3,000 feet and 70% above 3,000 feet. Aircraft ground operations and low-altitude flights produce emissions as described in Table 2.1. In addition, methane is not produced from aircraft flights. Still, it is generated by the GSE and APUs.[11]

***Aircraft Noise***

Continuing with aviation environmental impacts, noise is another adverse result of aviation operations. Noise pollution is the most frequent source of friction between an airport and its neighbouring communities. It can come from airports and their operations and aircraft engines, affecting the local areas and communities nearby. ICAO established in its 33rd Assembly Session in 2001 the policy on aircraft noise through the Balanced Approach to Aircraft Noise Management.[12] Even though ICAO leads in reducing noise and introducing abatement measures for noise from aircraft and airport operations, EASA and FAA propose measures and schemes for that exact purpose. Noise contours measure the airport noise and are modelled based on runway dimensions and aircraft operations, considering nighttime operations. The FAA uses average twenty-four-hour noise measurements to identify noise levels in the United States. This results in a day-night level (DNL) of noise recorded.

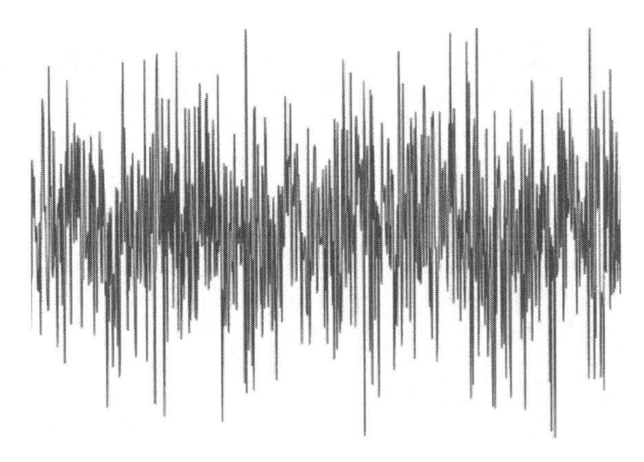

*Figure 2.2* Noise graph

> In an area, the daily average recorded noise level in an airport is at 65 decibels (dB). This is equivalent to a blender or a garbage disposal. An aircraft passing overhead is about 90dB, about the level of concert. When all the levels of noise in a day are averaged – including periods of silence – the average level is a more moderate DNL value.

There are two primary noise sources from an aircraft: drag or resistance around the various parts of a plane and the engine. The airframe generates noise, mainly as flaps, slats and wheels are deployed while landing, increasing the surface area exposed to the air. Studies have found that once the landing gear has been deployed, the noise experienced on the ground can increase by up to 10 decibels (dB). Because of the logarithmic scale of decibels, this is equivalent to doubling the noise. The most significant contributor to aircraft noise, however, is the engine. A jet engine pulls in air through a fan and compresses that air through a series of smaller-bladed fans until it is dense and highly flammable by injecting fuel. When jet fuel enters the

*Figure 2.3* Airplane turbine

compressed air and is burned, the rapidly expanding air is forced out at the rear part of the engine, turning the turbines that make the fans rotate, propelling the engine and, thus, the aircraft forward.

The fans generate noise, as does the engine's exhaust. Engine noise dominates on take-off and cruising, whereas noise from the airframe may be most noticeable upon landing. Many measures to reduce noise can enhance engine efficiency. For example, one of the primary jet engine improvements of the 1970s is the high-bypass turbofan engine. It allows most of the air it pulls to inflow around the compression and combustion areas rather than through them, providing some insulation from the noise generated within. However, some noise improvements can increase drag and slow the plane down, making them less likely to be adopted by aircraft manufacturers. While propeller-driven engines are not entirely quiet, there was a significant jump in noise when propeller planes started to be replaced with the first jets in the 1960s. Pressure from local communities and governments led to noise certification in the United States in 1968 through the Aircraft Noise Abatement Act. This meant that aircraft manufacturers had to meet noise standards before their planes could be allowed into service, the same way they already had to meet safety standards.[13, 14]

FAA follows the ICAO guidelines and has set specific noise specification standards. The FAA advisory circular on the noise levels for U.S. certified and foreign aircraft provides noise level data for aircraft certificated under 14 CFR 36. It categorises aircraft based on their engines' noise levels.[15] All aircraft with an airworthiness certificate in the United States must comply with the noise standard requirements and have a noise certification. The noise certification aims to ensure that the most current technology for noise reduction is applied in aircraft design. The purpose of the FAA is to continuously monitor and determine if a new standard is needed to keep the noise reduction levels appropriate for local communities. Thus, the FAA regulates the maximum noise level an individual civil aircraft can emit through specific noise certification standards. These standards designate changes in maximum noise level requirements by 'stage' designation. FAA civil jet aircraft has five noise standard stages, equivalent to the ICAO noise standards. Stage 1 is the loudest stage 5 is the quietest. Stage 5 is the currently acceptable level for turboprop and jet aircraft.[15]

Noise is an unwanted sound. Sounds that are pleasant in a certain volume can be annoying at a higher volume. The decibel is measured in the logarithmic scale, and it is the unit measuring the intensity of the sound.

The EU Directive 2002/49/EC includes the assessment and management of environmental noise (the Environmental Noise Directive – END). This directive is the leading EU document to identify noise pollution levels and to initiate the necessary actions from the EU Member States. The Environmental Noise Directive focuses on determining the exposure to environmental noise, ensuring that the information on environmental noise and its effects are accessible and available to the public so that measures can be developed to prevent and reduce environmental noise where necessary. The directive addresses human exposure, particularly in built-up areas, residences, schools, hospitals and other noise-sensitive buildings and areas. It does not apply to noise from domestic activities or created by neighbours, noise at the workplace means of transport or military activities in military areas. In August 2012, Regulation 746/2012 came into force, laying down implementing rules for the airworthiness and environmental certification of aircraft and related products, parts and appliances, as well as for the certification of design and production organisations. The new regulation contains the EASA certification specification CS-36, which applies to EASA Part 21 design requirements for noise specifications.[16]

## Market-Based Measures

Market-based measures are economic measures necessary to support the aviation sector's decarbonisation. They play a crucial role in supporting other decarbonisation solutions for the aviation industry until the sector is ready to rely only on its emission reduction measures. For commercial airlines, there are two types of economic measures: a Cap-and-Trade System, like the European Union Emissions Trading Scheme (EU-ETS), and a carbon offsetting, like ICAO Carbon Offsetting Reduction Scheme in International Airlines (CORSIA).[17]

The EU-ETS works under the 'cap-and-trade' principle. 'Cap' refers to the total amount of various greenhouse gas emissions generated by several types of installations included in the scheme. This set cap is lowered over time, so total emissions fall. Within the cap, installations within the scheme follow the cap and buy or receive emissions allowances to trade with one another as needed. The limit on the total number of available allowances ensures they have a value.

### *European Union Emissions Trading Scheme (EU-ETS)*

The European Union Emissions Trading Scheme (EU-ETS) applies to all EU countries, Iceland, Liechtenstein and Norway. Set up in 2005, EU-ETS is the world's first international emissions trading system. EU-ETS covers around 45% of the EU's industrial activities, and it aims to restrict emissions from more than 11,000 installations, such as power stations, industrial plants and airlines operating between the countries. The EU-ETS focuses on emissions that can be measured, reported and verified. Carbon dioxide (CO2) is calculated from power and heat generation, oil refineries, production of iron, aluminium, metals and steel, cement, lime, glass, ceramics, paper, acids and organic chemicals, among others. Nitrous oxide (N2O) is calculated from activities that include the production of nitric, adipic and glyoxylic acids and glyoxal and perfluorocarbons (PFCs) from aluminium production.[18] The scheme includes several types of industries and activities that are responsible for polluting emissions. Commercial aviation could not be an exception from that list. For the aviation sector, the EU-ETS applies only to flights between airports of the European Economic Area. For 2013–2020 the aircraft operators' total allowances were limited to 95% of the average emission of 2004–2006 called 'cap'. The allowances allocated to aircraft operators are valid only within the aviation sector; however, aircraft operators may purchase additional permits from the Joint Implementation and the Clean Development Mechanisms (CDM) for up to 1.5% of the number of allowances per year.[13]

*Table 2.2* The four phases of EU-ETS[18]

| *Phase 1 (2005–2007)* | *Phase 2 (2008–2012)* | *Phase 3 (2013–2020)* | *Phase 4 (2021–2030)* |
|---|---|---|---|
| **A three-year pilot of 'learning by doing' to prepare for phase 2, when the EU-ETS would need to function effectively to help the EU meet its Kyoto targets.** | It coincided with the first commitment period of the Kyoto Protocol, where the countries in the EU-ETS had concrete emissions reduction targets to meet. | Significant changes took place, from the other 2 phases. An actual implementation phase, with more sectors added. | The legislative framework of the EU-ETS for its next trading period (phase 4) was revised in early 2018 to enable it to achieve the EU's 2030 emission reduction targets, as part of the EU's contribution to the Paris Agreement. |

*Note:* The four phases of EU-ETS

In 2012, the EU-ETS started to measure CO2 emissions from aviation. Under that scheme, all Europe-operating airlines must monitor, report and verify their emissions and submit allowances against them. They receive tradable allowances covering a certain level of their annual emissions flights. EU-ETS has contributed to reduced emissions from the aviation sector, around 17 million tonnes per year, with compliance covering over 99.5% of actual emissions. Even though EU-ETS implementation has been exemplary and, for some time, the only market-based measure to monitor and reduce aviation emissions, the introduction of the ICAO CORSIA extended the EU-ETS scope to almost all international flights and all IATA members.

The Clean Development Mechanism allows a country with an emission reduction or emission limitation commitment to build an emission reduction project in a developing country. The project can earn Certified Emissions Reduction (CER) credits, and each credit is equivalent to one tonne of CO2 emissions, which can contribute towards meeting Kyoto targets.[19]

**What Is the Joint Implementation Mechanism?**

The Joint Implementation (JI), like the CDM, is a project-based mechanism under the Kyoto Protocol. It is limited to transactions between countries that have commitments to limit or reduce their GHG emissions under the protocol. The goal of the program is to increase market efficiency. It allows industrialised countries to invest in GHG abatement projects to another industrialised country.[20]

***ICAO CORSIA***

The Kyoto Protocol was signed in 1997 to extend the United Nations Convention on Climate Change to reduce greenhouse gas emissions.[21] The protocol included binding targets for all developed countries – Annex I countries. The Kyoto Protocol and the Paris agreement did not include specific reduction emission targets for aviation.[22] However, ICAO has been working on possible policy measures that would apply to aviation reduction emissions. The process was long and slow. At the 37th ICAO Assembly, the Carbon Neutral Growth (CNG) Goal was agreed to set that the aviation net carbon footprint in any year after 2020 should remain below the baseline year (2020) emission levels. In the 38th ICAO Assembly, this goal was reaffirmed and led to the CORSIA market-based scheme's development.[22] The Carbon Offsetting and Reduction Scheme for International Aviation (CORSIA) addresses the increase in total CO2 emissions from international aviation above 2020 levels. In October 2016, during the 39th ICAO Assembly, CORSIA was agreed upon and set into force for 2020.

From January 1, 2019, all carriers must report their CO2 emissions annually. It is forecasted that CORSIA will mitigate around 2.5 billion tonnes of CO2 and generate over USD 40 billion in climate finance between 2021 and 2035. CORSIA is an emission-offsetting market-based measure that applies to operators that fly internationally and produce more than 10,000 tonnes of CO2 emissions per year. However, many operators are below that threshold and are outside the scope of CORSIA entirely. Domestic flights are also outside the scope of CORSIA. Operators above the emissions threshold will need to monitor their emissions annually and buy emissions offsets for growth beyond 2020 levels by the end of the first compliance cycle (2021–2023). CORSIA is the sole market-based measure for international aviation emissions since it ensures a fair playing field for all operators globally.

In 2019, seventy-four countries, representing 75.96% of international commercial aviation, have agreed to participate in CORSIA.[23] Table 2.3 shows the flights included and excluded in the CORSIA Scheme.

CORSIA has three phases, and each phase lasts for a specific period.

- Voluntary participation: Pilot Phase (2021–2023).
- First Phase (2024–2026).
- Participation based on the 2018 Revenue Tonne Kilometres (RTK) data: Second Phase (2027–2035).

All members are encouraged to participate in the pilot and first phase of CORSIA. For the second phase, the participation criteria are 90% of global RTK or 0.5% of RTK. International flights between State pairs, including Least Developed Countries (LDCs), Small Island Developing States (SIDS) and Landlocked Developing Countries (LLDCs), are only within the applicability scope of the offsetting requirements if the state decides voluntarily to participate.

*Table 2.3* ICAO CORSIA flight characteristics

| *Included* | *Excluded* |
|---|---|
| International flights only | Domestic flights |
| Civil flights including:<br>– Scheduled flights<br>– Non-schedule flights<br>– Cargo<br>– Business aviation<br>– General aviation | Flights for:<br>– Head of states flights<br>– Military<br>– Customs and police<br>– Humanitarian, firefighting, medical |
| Airplanes with MTOM > 5,700 kg | Airplanes with MTOM ≤ 5,700 kg |
| Operators with annual CO2 emissions> 10k tonnes | Operators with annual CO2 emissions ≤10k tonnes |

*Note:* ICAO CORSIA flight characteristics[23]

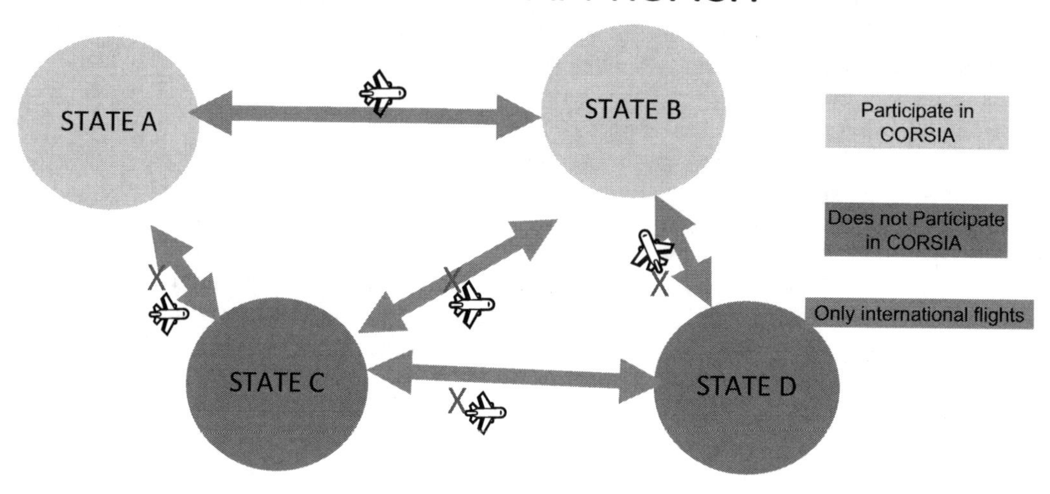


*Figure 2.4* CORSIA-route based

> When a flight is diverted from its original destination and it is an international flight, then according to the CORSIA offsetting requirements, the landing, even though it was not planned, the aircraft operator still will require to offset the emissions of such flights.[23]

The two schemes presented in this chapter, EU-ETS and CORSIA, address only one aspect of aviation's environmental effects, emissions. These market-based measures apply to commercial air operators and concern European domestic or international flights. However, authorities such as ICAO, FAA and EASA do not only focus on emissions from air operations. A series of plans and schemes apply to other activities of the aviation industry. For example, the FAA initiatives for environmental impact assessment and impact reduction in airports, the EASA certification specifications for noise and emissions reduction levels in aircraft design and engines and many more. Thus, it is essential to look at the aviation sector's whole picture. The holistic approach to sustainability must include all the industry's sectors. In the following sections, we will explore some essential topics supported by the leading aviation regulatory authorities like FAA and EASA.

***FAA Environmental Policy***

Federal Aviation Administration is the state organisation for managing all air transportation in the United States. Air traffic management, public protection during space launches, airport safety and inspections, design standards for airports, construction and operation, flight inspection standards and advancing satellite and navigation technology are some of the FAA's responsibilities.[24] In addition, FAA is deeply interested in and responsible for the environment and the effects of the aviation activities. The organisation has a set of rules and regulations with different scopes, activities and requirements. The National Environmental Policy Act (NEPA), the Clean Air Act (CAA) and other environmental regulations and directives control aviation emissions and air quality. The Clean Air Act (CAA) is the primary air quality law in the U.S.A. The CAA sets the Environmental Protection Agency (EPA) as the leading agency responsible for controlling air quality standards and developing regulations to meet those standards. EPA sets standards for air quality in urban areas. Under the CAA, EPA sets the limits for NO2, SO2, PM2.5, CO, O3 and lead (Pb).

Regarding aircraft engines and emissions, the CAA requires EPA to consult and take decisions with FAA. The FAA can enforce EPA's aircraft engine emissions standards through its certification regulations. Also, FAA ensures that these regulations do not affect aircraft safety or operational requirements. There are more than 60 standards for design in aircraft engines, construction materials, instruments and safety.[25] In addition, there are standards for the Fuel Venting and Exhaust Emission Requirements for Turbine Engine Powered Airplanes. These standards specify compliance with EPA's aircraft exhaust emission standards. The 1969 National Environmental Policy Act (NEPA)[26] requires federal agencies to consider the environmental impacts of their operations, which could include grants, loans, leases, permits and approval of plans or projects. Most airport construction projects follow the NEPA requirements under FAA's funding or approval. FAA has standards for air quality, emissions, engine exhaust fumes and design. Also, it provides a series of programs and actions for airports. Airports can apply low-cost energy efficiency measures like improving building insulation. These kinds of measures, at the same time, serve as GHG emissions and mitigation costs actions. For airport facilities, measures such as purchasing renewable energy and installation of renewable energy systems (given their compatibility with airport operations) or any action regarding energy consumption reduction, heating efficiency monitor, ventilation and cooling systems and purchasing low or zero-emission vehicles and GSE are also necessary for carbon emissions reduction.[27]

**VALE – Voluntary Airport Low Emissions Program**

The VALE is a program supported by the Clean Air Act developed in 2004. It is a program created to support airport stakeholders in reducing airport emissions. It provides funding for low-emission vehicles, stations for recharging and refuelling, gate electrification and other actions to improve airport air quality. Until September 2020, the VALE program funded 121 projects in fifty-six airports, and it is expected to reduce 1.370 tonnes per year until 2025.[28]

***EASA Environmental Initiatives***

The European Aviation Safety Agency (EASA) is the leading authority for all European Member States regarding aviation safety and operations. It publishes all regulations and directives for aviation operations, safety, maintenance, licenses, manufacturing design and certifications. As expected, EASA also has a set of regulations and rules to support the environmental effects of aviation activities. The basic EASA regulation EU 2018/1139 includes for the first time in Article 9 the requirements for environmental protection in aviation. This new article sets the environmental requirements for air quality and noise levels, including engine and propeller design and alignment with the ICAO CORSIA standards and the EU-ETS requirements.[29] Under this context, EASA has developed a series of standards to achieve its environmental objectives at the lower costs possible. EASA, with its actions and regulations, contributes to the ICAO Committee on Aviation Environmental Protection (CAEP). The ICAO CAEP develops and maintains the international standards for aircraft noise, $CO_2$ emissions, fuel venting and aircraft engine emissions, such as oxides of nitrogen (NOx), unburned hydrocarbons (HC), carbon monoxide (CO) and smoke and non-volatile particulate matter (nvPM). The latest environmental standards are included in Article 9(2) and Article 19(3) of the EASA Basic Regulation, Annex II (Part-21) of the Implementing Regulation and the Certification Specifications of CS-34 (emissions), CS-36 (noise) and CS-CO2 ($CO_2$ Emissions).[30]

*EASA Part 21: Certification Specification 34 and 36*

Certification Specifications (CS) 34 and 36 are parts of the initial aircraft design and manufacturing airworthiness regulation called Part 21. The initial issues of CS 34 and 36, published in September 2003, are developed based on the certification specifications and guidance materials of the ICAO Annex 16 documents and their annexes. CS 34 for aircraft engine emissions and fuel venting and CS 36 for aircraft noise are part of the implementing rules for aircraft engine emissions in EASA Part 21: Airworthiness and Environmental Certification. Paragraph Part 21A.18(c) provides both CS 34 and 36 acceptable means to show compliance with engine emissions and noise levels, respectively.[31, 32] Therefore, aircraft must be designed according to the noise and emissions levels mandated by the regulation and the certification specifications. As the wording indicates, the aircraft will be accompanied by the appropriate certifications, validating the regulatory compliance of their design. EASA certification noise levels are part of the aircraft certification process. All noise standards followed by EASA satisfy the ICAO Annex 16, Volume I. This annex is the baseline for all aircraft noise levels against which all National Aviation Authorities can issue individual noise certificates in the European Union.[33] EASA publishes in a database all the approved type certificate configuration data for noise levels and all supplemental certificates. In that manner, aircraft can demonstrate compliance with the specific standard and all Member States–approved repairs and modifications that are pending approval from EASA.[33] In that way, there is transparency between the aircraft designers, manufacturers, competent authorities and air operators. Moreover, all the aforementioned stakeholders can be facilitated when they approve a noise certification process, initial or amendment.

**What is the 'Certification Specification'?**

Certification specifications are agency measures used to comply with the Basic Regulation. They can be airworthiness codes used for technical interpretation of the Basic Regulation and for achieving the necessary airworthiness requirements. They can also be acceptable means of compliance; they are standards to comply with the regulation. However, they are not the only means to demonstrate compliance.[31]

Under a similar context, ICAO has set up an engine emissions databank. The ICAO aircraft engine emissions databank contains information on exhaust emissions of aircraft engines, measured according to the procedures in ICAO Annex 16, Volume II and, where noted, certified by the States of Design according to their national regulations. For the certification process against the specified emissions standards, engine manufacturers submit their data to a certificating authority responsible for the approval. Once the data are approved, EASA can add these data and specifications to the ICAO aircraft engine emissions databank.[34]

'Fit for 55' is the EU package including the upcoming plans for emissions reduction by at least 55% by 2030. The ultimate aim is for Europe to become the world's first climate-neutral continent by 2050. The proposals will affect a range of sectors as well as aviation, with the following three as the primary driving documents:

- the EU Greenhouse Gas Emissions Trading System (EU-ETS);
- the ReFuelEU Aviation – the uptake of sustainable fuels; and
- the revision of the Energy Taxation Directive introducing a tax on fuel for business and leisure flights.

The proposals and revisions of the EU-ETS mechanism foresees from 2024 to 2026 a progressive phase-out of the free allowances distributed to aircraft operators by 25%, 50% and 75% respectively and a complete phase-out of allowances from 2027 onwards. To meet the more stringent 2030 emissions target, the EU Commission proposes reducing the emissions cap by 4.2% annually instead of the current 2.2%. The EU-ETS would continue to apply to intra-EEA flights and flights to the U.K. and Switzerland, exempting those flights from the CORSIA Scheme. For other international flights, EU operators would be obliged to apply CORSIA. European Commission proposes as part of the 'Fit for 55' a legislation that will support the uptake of the Sustainable Aviation Fuels, the ReFuelEU Aviation. If ReFuel for Aviation is approved, then EASA will introduce control and monitoring mechanisms for air operators and fuel suppliers, assuring they are following the obligations of the reporting requirements included in the ReFuel for Aviation. The ReFuelEU Aviation initiative aims to create a set of rules at a European level, to maintain a competitive level playing field and to increase the uptake of SAF by operators and the increase distribution at Union airports. Europe aims to become the first climate-neutral continent by 2050 and to reduce emissions by 55% in 2030.[35, 36]

## Conclusion

This chapter presented some of the essential components that support the application of environmental sustainability in the aviation industry – and the aviation industry's environmental effects. The starting point is always the effects of aviation operations and the regulatory requirements for reducing these effects. FAA and EASA environmental regulations were described, along with the ICAO CORSIA and EU-ETS emission schemes. EASA and FAA have environmental initiatives and regulations that apply to the commercial aviation. However, they do not differ in their overall objectives. Regulatory requirements concerning reduced emissions and specifications for noise reduction are also a significant part of supporting environmental sustainability. The reduction of aviation environmental effects comes from both the operations and the design phases. New aircraft must follow design standards to reduce emissions and noise. At the same time, air operations must apply existing emission schemes and promote Sustainable Aviation Fuels (SAF) use. Finally, it is noteworthy that under the umbrella of environmental sustainability, environmental and energy management systems also play an essential role – a topic that we will examine in detail in Chapter 9.

## Key Points to Remember

- Environmental sustainability requires first to identify the effects and the sources that create these effects, then to investigate the existent environmental regulation and schemes.

- FAA and EASA, along with ICAO, propose extensive series and initiatives to support environmental sustainability in the aviation industry.
- Aviation environmental impacts do not occur only from aircraft flights. They can occur while aircraft are also on the ground. APU, GSE and GPU are also emission sources.
- Noise is also an environmental impact from aviation. FAA follows the ICAO noise measures and has set specific noise certification standards.
- There are market-based measures to control emissions in aviation.
- The EU has the Cap-and-Trade System. A cap is set on the total amount of certain greenhouse gas emissions from installations' emissions. The EU-ETS refers to all industrial activities that emit greenhouse gases.
- It is forecasted that CORSIA will mitigate around 2.5 billion tonnes of CO2 and generate over USD 40 billion in climate finance between 2021 and 2035.
- CORSIA is an emissions offsetting market-based measure that applies to operators that fly internationally and produce more than 10,000 tonnes of CO2 emissions.
- The FAA has a set of rules and regulations, with different scope, activities and requirements. The National Environmental Policy Act (NEPA), the Clean Air Act (CAA) and other environmental regulation and directives control aviation emissions and air quality.
- The Clean Air Act (CAA) is the primary air quality law in the U.S.A. The CAA sets the Environmental Protection Agency (EPA) as the main agency responsible to control air quality standards and develop regulations to meet those standards.
- The VALE is a program supported by the Clean Air Act, developed in 2004. It is a program created to support airport stakeholders in reducing airport emissions.
- EASA also has a set of regulations and rules to support the environmental effects of aviation activities. The basic EASA regulation EU 2018/1139, including for the first time in Article 9 the requirements for environmental protection. This new article sets the environmental requirements for air quality and noise levels, including engine and propeller design and alignment with the ICAO CORSIA standards and the EU-ETS requirements.
- Certification Specifications (CS) 34 and 36 are parts of the initial aircraft design and manufacturing airworthiness regulation called Part 21. The initial issues of CS 34 and 36, published in September 2003, are developed based on the certification specifications and guidance materials of the ICAO Annex 16 documents and their annexes.
- CS 34 for aircraft engine emissions and fuel venting and CS 36 for aircraft noise are part of the implementing rules for aircraft engine emissions in EASA Part 21: Airworthiness and Environmental Certification.
- ICAO has set up an engine emissions databank. The ICAO aircraft engine emissions databank contains information on exhaust emissions of aircraft engines, measured according to the procedures in ICAO Annex 16, Volume II and, where noted, certified by the States of Design.
- 'Fit for 55' is the EU package including the upcoming plans for emissions reduction by at least 55% by 2030.
- The ultimate aim is for Europe to become the world's first climate-neutral continent by 2050. The proposals will affect a range of sectors as well as aviation, with the following three as the primary driving documents:
  - the EU Greenhouse Gas Emissions Trading System (EU-ETS);
  - the ReFuelEU Aviation – the uptake of sustainable fuels; and
  - the revision of the Energy Taxation Directive introducing a tax on fuel for business and leisure flights.

*Table 2.4* Acronym rundown

| | |
|---|---|
| AIRE | Atlantic Interoperability Initiative to Reduce Emissions |
| APU | Auxiliary Power Unit |
| CAA | Civil Aviation Authority |
| CAEP | Committee on Aviation Environmental Protection |
| CDM | Clean Development Mechanism |
| CER | Certified Emissions Reduction |
| CNG | Carbon Neutral Growth |
| CORSIA | Carbon Offset and Reduction Scheme for International Aviation |
| CS | Certification Specification |
| EC | European Commission |
| EPA | Environmental Protection Agency |
| EU-ETS | European Union Emissions Trading Scheme |
| GPU | Ground Power Unit |
| GSE | Ground Support Equipment |
| HC | Hydro-carbons |
| JI | Joint Implementation |
| ICAO | International Civil Aviation Organization |
| IPCC | Intergovernmental Panel on Climate Change |
| LDC | Least Developed Countries (SIDS and LLDCs) |
| LLDC | Landlocked Developing Countries |
| NEPA | National Environmental Policy Act |
| nvPM | non-volatile particulate matter |
| PFC | Perfluorocarbons |
| RTK | Revenue Tonne Kilometres |
| SIDS | Small Island Developing States |
| WMO | Meteorological Organization |
| UNEP | United Nations Environmental Programme |

**Chapter Review Questions**

2.1 Explain what environmental sustainability means.

2.2 Give a description on what environmental sustainability is for the aviation industry.

2.3 Do you believe that emission schemes are sufficient to support environmental sustainability?

2.4 What is the role of International Civil Aviation Organization to aviation's environmental sustainability?

2.5 Identify the requirements of an airline to be included in the ICAO CORSIA Scheme.

2.6 What are the flights excluded from ICAO CORSIA?

2.7 Identify three types of aviation emissions. Explain their source and their impact to the natural environment.

2.8 Extend your research and identify the role of APU in the noise problem in airports. Research on the existing literature and identify a particular case.

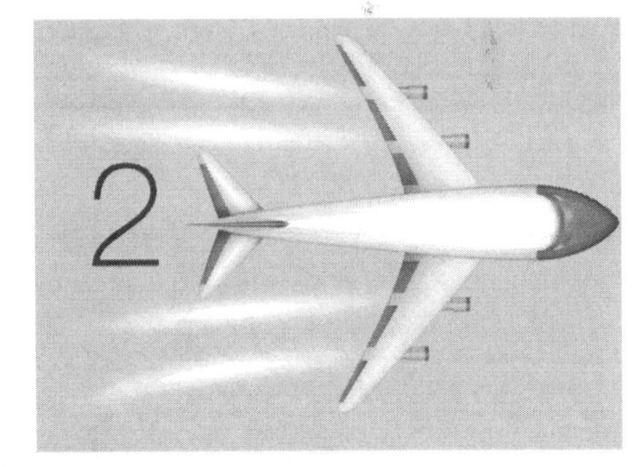


*Figure 2.5* Airplane no. 2 pointing right

2.9 What are some initiatives applied from airports to reduce noise?
2.10 What is the 'Fit for 55' package?
2.11 What are the EASA certification specifications 34 and 36?

Reduced Vertical Separation Minima (RVSM) is a program created by the International Civil Aviation Organization (ICAO) in 1982. The initial purpose of this program is to allow an increase in available flight levels for aircraft to operate in, reducing the threat of having multiple aircraft conflicting in certain locations. Additionally, it enables airlines to enhance their operational optimisation and minimise air traffic controllers' workload when traffic gets heavy. From 1997 until 2005, RVSM was implemented globally, permitting airlines to take full advantage of its benefits. Before RVSM, Vertical Separation Minima (VSM) only allowed aircraft to have 2,000 feet separation between each aircraft. RVSM has also allowed airlines to utilize wind prediction models to optimize their operational flight plans, allowing them to operate in environments where less turbulence is predicted or less headwind is projected. In theory, flight level capacity is doubled in RVSM airspace, meaning higher air traffic capacity. Since there are more flight levels available, it can help pilots fly closer to optimum altitude to save fuel, and they have a better chance of avoiding turbulence on certain flight levels. This allows a higher passenger comfort as well as a higher operational efficiency as on-time performance (OTP) requirements are met. After you have read about the RVSM, extend further your research and answer the following questions.

***Case Study Questions***

2.1 What is the flight level separation due to the RVSM?
2.2 How does RVSM benefit environmental sustainability and why?
2.3 What are some benefits that airlines can accomplish by following the RVSM program?
2.4 What is the effect of RVSM in a flight's fuel consumption and greenhouse gas emissions?
2.5 How does RVSM support all three pillars of sustainability?
2.6 Can RVSM support an airline's sustainability goals and how?

## References

[1] The Intergovernmental Panel on Climate Change. (2018, October 8). *Summary for policymakers of special report on global warming of 1.5Â°C approved by governments*. Retrieved February 18, 2022, from www.ipcc.ch/2018/10/08/summary-for-policymakers-of-ipcc-special-report-on-global-warming-of-1–5c-approved-by-governments/
[2] European Aviation Safety Agency. (n.d.). *Aviation environmental impacts | European aviation environmental report*. Retrieved February 18, 2022, from www.easa.europa.eu/eaer/climate-change/aviation-environmental-impacts
[3] SkyBrary. (2021, May 26). *Auxiliary power unit (APU)*. SKYbrary Aviation Safety. Retrieved February 18, 2022, from https://skybrary.aero/articles/auxiliary-power-unit-apu
[4] SkyBrary. (2021, January 8). *Ground power unit (GPU)*. SKYbrary Aviation Safety. Retrieved February 18, 2022, from https://skybrary.aero/articles/ground-power-unit-gpu
[5] Federal Aviation Agency. (2013). *Federal aviation administration, DOT*. Discover Government Information. Retrieved February 18, 2022, from www.govinfo.gov/content/pkg/CFR-2013-title14-vol3/pdf/CFR-2013-title14-vol3-sec158-3.pdf

[6] ICAO. (n.d.). *Aircraft engine emissions*. ICAO Environment. Retrieved February 18, 2022, from www.icao.int/environmental-protection/pages/aircraft-engine-emissions.aspx
[7] EASA. (2022). *Aviation environmental impacts | European aviation environmental report*. Retrieved February 18, 2022, from www.easa.europa.eu/eaer/climate-change/aviation-environmental-impacts
[8] NOAA. (n.d.). Climate forcing | NOAA climate.gov. *Climate.Gov*. Retrieved February 18, 2022, from www.climate.gov/maps-data/climate-data-primer/predicting-climate/climate-forcing
[9] FAA. (2015). *Aviation emissions, impacts and mitigation: A primer*. Retrieved February 18, 2022, from www.faa.gov/regulations_policies/policy_guidance/envir_policy/media/primer_jan2015.pdf
[10] NASA. (2008, November 17). *NASA – water vapor confirmed as major player in climate change*. Retrieved February 18, 2022, from www.nasa.gov/topics/earth/features/vapor_warming.html
[11] Environmental and Energy Study Institute (EESI). (2019, October 17). *Fact sheet | The growth in greenhouse gas emissions from commercial aviation (2019) | white papers | EESI*. Retrieved February 18, 2022, from www.eesi.org/papers/view/fact-sheet-the-growth-in-greenhouse-gas-emissions-from-commercial-aviation
[12] International Civil Aviation Organisation ICAO. (2013). *Resolutions, adapted by the assembly*. Assembly – 38th Session. Montreal. www.icao.int/Meetings/GLADs-2015/Documents/A38–18.pdf
[13] Scheelhaase, J., Maertens, S., Grimme, W., & Jung, M. (2018). EU ETS versus CORSIA – A critical assessment of two approaches to limit air transport's CO2 emissions by market-based measures. *Journal of Air Transport Management*, *67*, 55–62. https://doi.org/10.1016/j.jairtraman.2017.11.007
[14] Garrett, M. (2014). Noise pollution and airplanes. In *Encyclopedia of transportation: Social science and policy* (Vol. 1, pp. 993–994). SAGE Publications, Inc. https://dx.doi.org/10.4135/9781483346526.n351
[15] FAA. (2015b, September 14). *Aircraft noise levels & stages*. Retrieved February 18, 2022, from www.faa.gov/air_traffic/noise_emissions/noise_levels/
[16] European Union. (2002, July 18). *Directive 2002/49/EC of the European parliament and of the council of 25 June 2002 relating to the assessment and management of environmental noise*. European Union Law. Retrieved February 18, 2022, from https://eur-lex.europa.eu/LexUriServ/LexUriServ.do?uri=OJ:L:2002:189:0012:0025:EN:PDF
[17] European Region Airline Alliance. (n.d.). *Market-based measures | ERA*. Retrieved February 18, 2022, from www.eraa.org/policy/sustainable-aviation/market-based-measures
[18] European Union. (n.d.). *EU emissions trading system (EU ETS)*. Climate Action. Retrieved February 18, 2022, from https://ec.europa.eu/clima/eu-action/eu-emissions-trading-system-eu-ets_en
[19] United Nations for Climate Change. (n.d.). *The clean development mechanism*. Retrieved February 18, 2022, from https://unfccc.int/process-and-meetings/the-kyoto-protocol/mechanisms-under-the-kyoto-protocol/the-clean-development-mechanism
[20] Greenhouse Gas Management Institute & Stockholm Environment Institute. (n.d.). *Joint implementation*. Carbon Offset Guide. Retrieved February 18, 2022, from www.offsetguide.org/understanding-carbon-offsets/carbon-offset-
[21] United Nations. (1998). *Kyoto protocol to the united nations framework convention on climate change*. Retrieved February 18, 2022, from https://unfccc.int/resource/docs/convkp/kpeng.pdf
[22] Verifavia. (n.d.). *Greenhouse gas emissions (GHG) verification aviation, airports, shipping Verifavia*. Retrieved February 18, 2022, from www.verifavia.com/greenhouse-gas-verification/fq-why-were-inter nationaeuro-al-aviation-emissions-not-included-in-the-paris-agreement-at-cop-21–242.php
[23] ICAO. (2019). *Environmental technical manual*. ICAO. https://elibrary.icao.int/reader/234240/&returnUrl%3DaHR0cHM6Ly9lbGlicmFyeS5pY2FvLmludC9ob21lL3Byb2R1Y3QtZGV0YWlscy8yMzQyNDA%3D
[24] FAA. (2022, January 21). *FAA workforce*. Retrieved February 18, 2022, from www.faa.gov/jobs/who_we_are/
[25] Environmental Protection Agency. (2013, January 15). *National ambient air quality standards for particulate matter*. Federal Register. Retrieved February 18, 2022, from www.govinfo.gov/content/pkg/FR-2013-01-15/pdf/2012-30946.pdf
[26] NEPA. (2021). *National environmental policy act implementing regulations* (40 CFR Parts 1500–1508). https://ceq.doe.gov/docs/laws-regulations/nepa-implementing-regulations-desk-reference-2021.pdf
[27] FAA. (2022, February 17). *Airport carbon emissions reduction – airports*. Retrieved February 18, 2022, from www.faa.gov/airports/environmental/air_quality/carbon_emissions_reduction/

[28] FAA. (2021, November 4). *Voluntary airport low emissions program (VALE) – airports*. Retrieved February 18, 2022, from www.faa.gov/airports/environmental/vale/
[29] European Union. (2018). Regulation (EU) 2018/1139 of the European parliament- regulations. *Official Journal of the European Union*, *L*(212), 1–122. https://eur-lex.europa.eu/legal-content/EN/TXT/PDF/?uri=CELEX:32018R1139&from=EN
[30] EASA. (n.d.). *Smart environmental standards*. Retrieved February 18, 2022, from www.easa.europa.eu/domains/environment/smart-environmental-standards
[31] EASA. (2019, July 29). *Explanatory note*. Retrieved May 19, 2021, from www.easa.europa.eu/sites/default/files/dfu/expnote_cs34.pdf
[32] EASA. (2019, July 29). *Certification specifications, acceptable means of compliance and guidance material for aircraft noise*. Retrieved May 19, 2021, from www.easa.europa.eu/sites/default/files/dfu/CS-36_Amendment_5.pdf
[33] EASA. (2021). *Certification noise levels*. EASA. Retrieved February 18, 2022, from www.easa.europa.eu/domains/environment/easa-certification-noise-levels
[34] EASA. (2022). *ICAO aircraft engine emissions databank*. Retrieved February 18, 2022, from www.easa.europa.eu/domains/environment/icao-aircraft-engine-emissions-databank
[35] EASA. (2021, July 14). *European commission publishes 'Fit for 55' legislative package*. Retrieved February 18, 2022, from www.easa.europa.eu/newsroom-and-events/news/european-commission-publishes-fit-55-legislative-package
[36] Eurocontrol. (2021, October 27). *The EU's "fit for 55" package: What does it mean for aviation?* www.eurocontrol.int/article/eus-fit-55-package-what-does-it-mean-aviation

# 3 Economic Aspects of Aviation Sustainability

## Chapter Outcomes

At the end of this chapter, you will be able to do the following:

- Recognize the economic benefits of aviation and aerospace.
- Learn the three different types of economic benefits of aviation industry.
- Identify how aviation supports economic sustainability.
- Provide an analysis on aviation industry economic effect in other economic activities.
- Explain what linear and circular economy is.
- Identify the basic actions to create a circular economic model for the aviation industry.

## Introduction

*Figure 3.1* Aviation economy right arrow

It is essential to explain the economic benefits of aviation activities, aiming to identify what elements constitute economic sustainability and how it can become part of aviation and aerospace operations and enhance sustainability holistically. Aviation is an industry with worldwide economic contribution and is the most 'global' of all other sectors. The interaction of aviation and aerospace activities with other commercial activities offers many economic benefits. However, it is essential to create sustainable economies that one support another. We often hear the term sustainable economy or sustainable green economy. There is also the sustainable blue economy. The difference is not in colour. The difference is in focus. Identifying the elements that will create a sustainable economy for different sectors is essential. Why do we call an economy green or blue? Does a colour code connect with the economy and its activities? An economy, and especially the economy of the aviation and aerospace, should be sustainable, but after current conditions through volcano eruptions or, more recently, through COVID-19, it should also be resilient. Exploring the economic activity of the aviation and its economic benefits to the global economy and society must be in conjunction with protecting the environment. Economic

DOI: 10.4324/9781003251231-3

models and theories can be applied and adjusted to the aviation and aerospace ecosystem so that the industry can support sustainability.

## What Is Economic Sustainability?

*Figure 3.2* Economic sustainability right arrow

When we hear the word economy, our minds go to profit, economic development or simply to money. Economic sustainability as part of the sustainability 'umbrella' is a lot more than that. Economic sustainability refers to the long-term development of an organisation or system in a way that it will not affect the other two pillars, environment and society, or other relevant concepts of them. Economic growth and profit are vital elements for any organisation's viability. However, economic sustainability should always be in balance the other two pillars. Inevitably, economic growth and development can support the social aspect of sustainability, both internally and externally, of a business entity. Additionally, environmental management practices, waste management or energy efficiency investments can only be supported through a financially viable system. For the aviation industry, economic sustainability is based on the same foundations as all other economic entities – to have a viable growth to continue to operate and achieve benefits internally and externally but with equal consideration of the environment, the human element and the society.

> Economic sustainability in aviation means considering economic development and profit to support the industry's activities and people, respect the people, internally and externally, develop activities and use technologies which will protect and support the natural environment.

### *The Five Capitals Model*

The five capitals model explains the different sustainability principles under the economic concept of wealth creation or 'capital'. The model explains that each organisation must deliver its products and services under the five types of capital. A sustainable company must preserve and provide all the necessary assets to develop the five types of capital. Using this model for decision-making, business entities, like aviation-related ones, can get sustainable outcomes through their operations. The five capitals are the natural capital, the social capital, the human capital, the manufactured capital, and the financial capital. Everyone knows the word *capital*, which is directly associated with economic concepts, values and, in our case, the economic aspect of sustainability. However, each capital aims to provide wealth and, consequently, the necessary resources, as already said, to support the harmony among the three pillars in a system.

- **Natural Capital**: It refers to the environmental part of sustainability. It includes all the necessary resources, energy consumption, water and fossil fuels and all the natural processes. The waste and pollutants generated by these processes and consumption must be minimised, recycled or even eliminated, naturally if possible, and prevented from polluting the natural environment.
- **Social Capital**: It refers to all the activities and mechanisms that support people's communication and collaboration, with safe and respectful working conditions. Diversity, inclusion and ethics are some of the elements that should be inclusive to the social aspect of sustainability through the support of the social capital provided by the organisation.
- **Human Capital**: It refers to the people and the practical aspects that are necessary to support their work. Healthy working conditions, growth opportunities, trainings to develop skills and forward-thinking work standards are also part of the social sustainability, which is supported by the human capital too.
- **Manufactured Capital**: This capital includes tangible resources such as buildings, transport networks, tools, machines and computers, as capital should be flexible, innovative and apply operations to decrease resource usage and increase efficiency.
- **Financial Capital**: It refers basically to the currency needed to cover all the assets. It is basically money![1]

The financial capital covers the assets in the form of currency, which basically means money! In a very common sense, if a business is not profitable and doesn't make money, even if the other capitals are being well-managed, then definitely, it is not sustainable nor viable. The financial capital can cover many issues such as fair distribution of wealth and creating wealth in local communities within which a business unit operates. Taking into account also the other capitals, financial capital plays a critical role when determining the financial position to the market. Consequently, when the financial capital does not bring profit and growth to an organisation, it will affect the growth of the other capitals of the same organisation.

## Aviation Economic Benefits

The role of air transportation to the modern society and business world inevitably is that aviation connects people. It supports the transportation of people and goods and people facing health issues in remote areas. It is the most 'global' of all industries. Obviously, various workers, partners and stakeholders support these needs through air transportation. They all work together in a complex choreography of activities to maximise the industry's benefits and support aviation's sustainable growth by connecting more people and places more often.[2] Aviation is a sector with global economic development and input. Over 30% of all trade is sent by air, positioning aviation as a critical business factor worldwide. It supports national economies to become part of the global economy, offering direct benefits through its economic growth. Nevertheless, what does economic sustainability mean for the aviation? We must remember that economic growth is not synonymous to economic sustainability. For economic growth to be sustainable, the aviation sector must adopt sustainable economic practices. Therefore, it is essential defining and understanding what makes aviation a sustainable sector and how it can operate under economic sustainability.

Economic sustainability refers to practices that support long-term economic growth, not only without negatively affecting the other two pillars of sustainability – environment and society – but supporting them also to exist and continuously develop as parts of a sustainable system.

We should remember that economic sustainability is one of the three pillars of sustainability, and the sustainability core concept is the balanced approach that requires all three components to develop and exist equally. Hence, economic sustainability must be achieved and practised under that perspective. Accordingly, all other industries aiming to be sustainable should follow this basic principle. Aviation's economic benefits are comprised of three main categories: direct, indirect and induced. The economic benefits to support aviation sustainability must benefit also the social element involved. This means that we cannot discuss about an economically sustainable aviation system when only economic growth exists. The aviation social element must be benefited and supported from the sector's economic growth. Chapter 4 explains the social aspect of aviation sustainability and how it connects with the economic aspect of aviation activities.

The economic activity of aviation creates jobs that directly serve passengers at airlines, airports and air navigation providers. These jobs include check-in staff, ground handlers, on-site retail, catering facilities and cargo and baggage handlers. Aircraft and aircraft parts manufacturers are also considered significant contributors with direct economic benefits since they produce aircraft, engines and other aircraft technologies and equipment and require a highly-skilled and trained workforce. Air crew personnel, aviation maintenance engineers, organisations that provide the licenses for pilots, cabin crew and maintenance staff are also an essential part of the aviation economy.

Part of aviation's indirect economic benefits are the employment and economic activity from aviation fuel suppliers, constructors of airport facilities, suppliers of any sub-components needed in the aircraft, manufacturers, goods sold in the airport retail shops, call centres and IT facilities and services, among others.

**Induced Economic Benefits**[3]

The economic activity of the directly or indirectly employed in the aviation industry supports the creation of additional employment in other sectors. For example, banks, hotels, restaurants and other job positions are created from the spending of the indirect and direct workforce of the aviation sector.

***Employment***

Employment is a core element of the economy; it facilitates trade, supports the rapid growth of all business sectors and assists people in covering their personal, professional and consumer needs. These are some of the most apparent benefits someone can identify from employment and its support to an economy. Aviation is a broad sector that gives employment to many people across the globe. At a global level, the aviation supports 87.7 million jobs, directly or indirectly, linked simultaneously with the direct, indirect or induced financial impacts mentioned earlier. Different aviation professions comprise this number, including aviation operators, air operators' employees such as pilots, cabin crew, ground services and catering staff, maintenance staff and all managerial and administrative staff, aerospace and manufacturing employees, air traffic controllers, executives, etc.

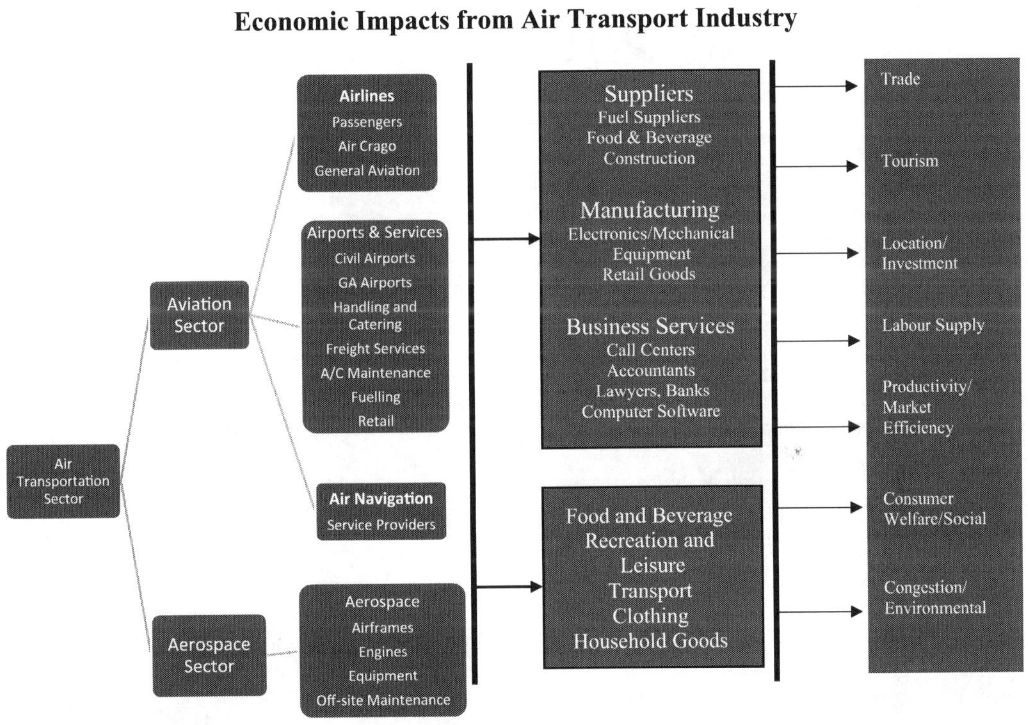

**Economic Impacts from Air Transport Industry**

Air Transportation Sector
Aviation Sector
Aerospace Sector
Airlines
Passengers
Air Crago
General Aviation
Airports & Services
Civil Airports
GA Airports
Handling and Catering
Freight Services
A/C Maintenance
Fuelling
Retail
Air Navigation
Service Providers
Aerospace
Airframes
Engines
Equipment
Off-site Maintenance
Suppliers
Fuel Suppliers
Food & Beverage
Construction
Manufacturing
Electronics/Mechanical
Equipment
Retail Goods
Business Services
Call Centers
Accountants
Lawyers, Banks
Computer Software
Food and Beverage
Recreation and Leisure
Transport
Clothing
Household Goods
Trade
Tourism
Location/ Investment
Labour Supply
Productivity/ Market Efficiency
Consumer Welfare/Social
Congestion/ Environmental

*Figure 3.3* Operational and economic activities in the air transportation industry[4, 5]

Figure 3.3 shows a classification of employment in direct, indirect and induced jobs. Direct jobs cover around 11.3 million people and are directly related to aviation activities. The share of each direct employment is shown in Figure 3.4. The aviation industry supports around 18.1 million indirect jobs by purchasing goods and services.

The employment type of indirect jobs includes fuel suppliers, construction companies, aircraft component suppliers, manufacturers that cover airports' needs and various business roles from call centres, administration, accounting, IT to technical staff and other supporting roles.

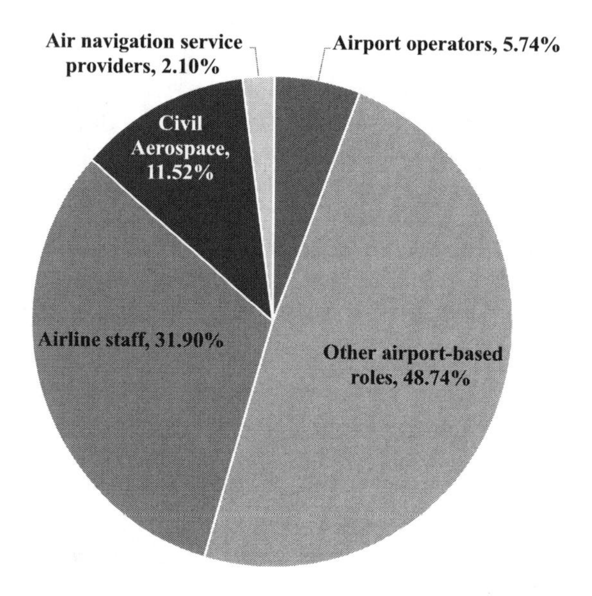


*Figure 3.4* Direct jobs employment share[5]

Similarly, air transport industry employees, direct and indirect, keep around 13.5 million induced jobs. This number of employees supports the purchase of goods and other services to cover their needs with their income.

***Connectivity, Tourism and Trade***

When we hear the word 'connectivity', we think of trips and remote destinations that we can reach with transportation appropriate for long journeys. However, connectivity does not have to do only with the arrival to a destination that is meant to be only for holidays; it also has to do with trade. Connectivity means connecting two or more different geographical points through the transportation of people or cargo, preserving certain quality conditions and characteristics. Everyone wants their parcel delivered, if not on time, at least as promised to arrive when purchased. Similarly, passengers expect to arrive at least, as their flight was scheduled, with their luggage waiting for them and not lost between airports and, needless to say, under safe conditions. There is a great network behind

*Figure 3.5* The globe

that simplistic explanation to achieve this successful transition from one point to another, transporting people or goods. This is the air transport network. Someone might wonder, though, how the air transport network links with sustainability. Well, the answer might not be so simple, but there is a strong connection between the role of the air transport network and economic sustainability.

*Figure 3.6* The air transportation network

Air transportation has a network comprised of scheduled airlines, aircraft in service and ready to fly when planned and airports facilitating flights in arrivals and departures. Inevitably, when there is good connectivity in destinations, economic benefits can grow and bring development in remote areas of having less accessibility to goods and services. Naturally, when connectivity is well-established, other industries can be supported. The connectivity between countries and regions creates an economic net that supports tourism, leisure industry and catering businesses, supporting local communities that rely on this type of economic activity. However, connectivity supports another critical area of economy, trade. Trade is linked to the transportation of goods produced in other places than where they are sold. Air transportation supports extensively this type of commercial activity, including freight and cargo that also assist economic growth and trade values. Nevertheless, we expect nowadays that trade, tourism and every element that supports connectivity to operate under the sustainability principles, aiming to meet the 17 SDGs and create a better, safer and long-lasting and resilient society.

Tourism is one of the most pleasant activities for people and relies in good connectivity. Travelling within their country of origin, or out of it, people can meet new cultures and explore globally known monuments and historic places. It is a fast-growing industry, and it will continue to be, with aviation's immense support. The connectivity brought by air transport is at the heart of tourism development, providing substantial economic benefits for all those involved in the tourism value chain. Approximately 1.4 billion tourists cross borders annually, over half of whom arrive at their destinations by air.[3] Therefore, there is an invisible connecting string between the two industries, and they are strongly interrelated and interconnected. Tourism gives the opportunity to people to rest and feel more relaxed while they are on holidays. Aviation supports tourism as it transports people in closer or very distant areas, from people's origins. Social sustainability has as its core element, the society and humans. An industry that can support peoples' recreational activities, apart from business trips, offers a lot more than 'just a trip'. Going back again in 2020, the year where COVID-19 started spreading, we all observed that any kind of flight, apart from cargo and freight (however, even those where limited), was ceased due to the pandemic. Holiday trips with airplanes were ceased as an unnecessary activity and as a measure to prevent the spread of the pandemic. Because of that, people couldn't travel for leisure and see their families and friends. Everyone's psychology was severely affected because of all the domino effects from COVID-19. At the same time, due to air transportation activity for tourism activities, there are many economic benefits, directly and indirectly. The economic activity from tourism and air transportation supports areas and regions that rely heavily on the tourism industry. Employment is boosted, and people can have a good quality of life covering their daily needs and much more. People that have employment and are in islands or other remote destinations can enjoy not only economic benefits but also benefits to their societies.

Aviation carries 1% of the world's transported weight; in terms of money, this value is about 35%, which shows the necessity of aviation in global logistics.

On the other side of connectivity, commercial activities and transportation of goods rely heavily on aviation. With the words 'freight' or 'cargo', a wide range of products and goods are transferred, facilitating people's lives economically, emotionally and socially. Final products are delivered to everyone who makes an order either for personal use or for their business, factory, machinery or assembly facility. Mail of documents, letters or packages is also part of the goods that must be transported and serve financial and social purposes. Other products include food, medicine, clothes, hardware accessories and equipment that need quick transportation. This is where air transport makes the difference due to its short travel time compared to other means. Life stock and pharmaceuticals are also prone to long transportation times, as they must be kept under special preservation conditions. Subsequently, we see only the endpoint of aviation operations when we talk about air transportation. However, behind that endpoint is a broad and complex net of aviation and aerospace operations interlinked with solid economic bonds.

Nevertheless, what happens when this super-complex net ceases to interact and operate? How are economies affected? When answering these questions, aviation role in economic sustainability can be clearly explained. Economic sustainability for all aviation sectors means being able to function even in challenging conditions or even be prepared for disruptions so the interacting systems are not affected in a significant way. At the same time, an economically sustainable aviation system must offer the necessary resources to support the working force.

### *When Everything Stops*

The aviation sector serves greatly its core objective, which is the transportation of passengers and cargo. But what happens when everything stops? What happens when unforeseeable conditions mandate the cease of aviation operations for a short period or, even worse, for more extended ones? In 2010, the European airspace was suspended from flights for almost a week, affecting mainly the northwest area of Europe. It was a short but severe disruption of flights due to extreme safety concerns. Ten years later, in 2020, a more severe disorder happened. A cease of almost all air operations occurred due to the COVID-19 pandemic. The industry was greatly affected for more than two years, with slow steps to normality after 2021. Especially in the case of COVID-19, it was obvious that the aviation world was not prepared for such a turmoil. Many companies ceased major parts of their international offices, airlines dismissed hundreds of thousands of cabin crew members and maintenance technicians and airports could not sustain a large number of ground support personnel without any flights in the terminals.

*Figure 3.7* Eyjafjallajökull Volcano

On April 14, 2010, the Icelandic volcano Eyjafjallajökull (try to repeat it three times, quickly!) erupted, releasing into the atmosphere ash that blew over three kilometres, suspending most flights in Europe's airspace. Inevitably, even for such a short period, the modern world was seriously affected by the cease of aviation operations. But how has this disruption affected aviation in numbers?

- Around 10 million passengers were stranded, with over 100,000 flights being cancelled over five weeks, although the main shutdown was for almost a week.

- Almost 30% of the global air traffic capacity was affected.
- The hospitality sector was primarily affected by a global loss of visitors reaching around $1.6 billion.
- Spoilable goods and immediate production processes were significantly affected because of the trade disruption at the European and global levels.
- Following the first-week airspace shutdown, another 5,000 flights were cancelled.

The most significant hit has been seen in global trade from perishable goods, vegetables, flowers and just-in-time production components.[6]

For almost a decade, global air traffic was seeing ongoing growth until the COVID-19 pandemic hit. In 2020, and for nearly four months, all airports around the globe gradually ceased to operate. This crisis erased twenty-year passenger traffic and its growth almost overnight.

- In 2020, more than 5.9 billion passengers stopped travelling.
- The global passenger traffic declined by 63.2% compared to the projected baseline.
- The decline was 61% less than 2019 recorded global passenger traffic levels.
- Europe, the Middle East and Africa were the most impacted regions, and traffic fell 70%, 68.5% and 67.6%, respectively.
- Even though the Asia-Pacific region had a minor decline globally, with 55.5% from the projected levels and 53.7% compared to 2019, the area recorded the highest loss of traffic passengers – 1.9 billion passengers. After being hit first by COVID-19, it was the first to show signs of recovery due to the size and magnitude of China's market.
- After the April 2020 lockdown, the international passenger volume ended, slightly below 1 billion passengers, decreasing 73.7% compared to 2019.

**Aviation and Economic Circularity**

After the turmoil of COVID-19, or at least after the first months of a slow recovery, the industry needed a shift to a more resilient and sustainable way of operations. The effect of COVID-19 left millions of people professionally involved with the aviation and aerospace industry unemployed, with scarce chances of finding another job. Pilots, crew staff, maintenance technicians and engineers and airport staff found themselves unemployed within a week or days. The industry was unable to fight such an unexpected downturn, even though it had recently recovered from the 9/11 attack (2001), H1N (2009) and even the global economic crisis (2010). There was one way for the industry to recover, strengthen and remain resilient. Sustainability became the tool for recovery, and mainly, the economic sustainability pillar would have to support the other two pillars. The circular economy became a necessity rather than a 'new' trend to follow, as it was seen until 2020. The environmental, social and governance (ESG) model made a strong appearance, too, in the aviation business world. Aviation and aerospace, one of the greatest victims of COVID-19 in the global economy, drove the shift and part of the solution they needed at the time and for the future.

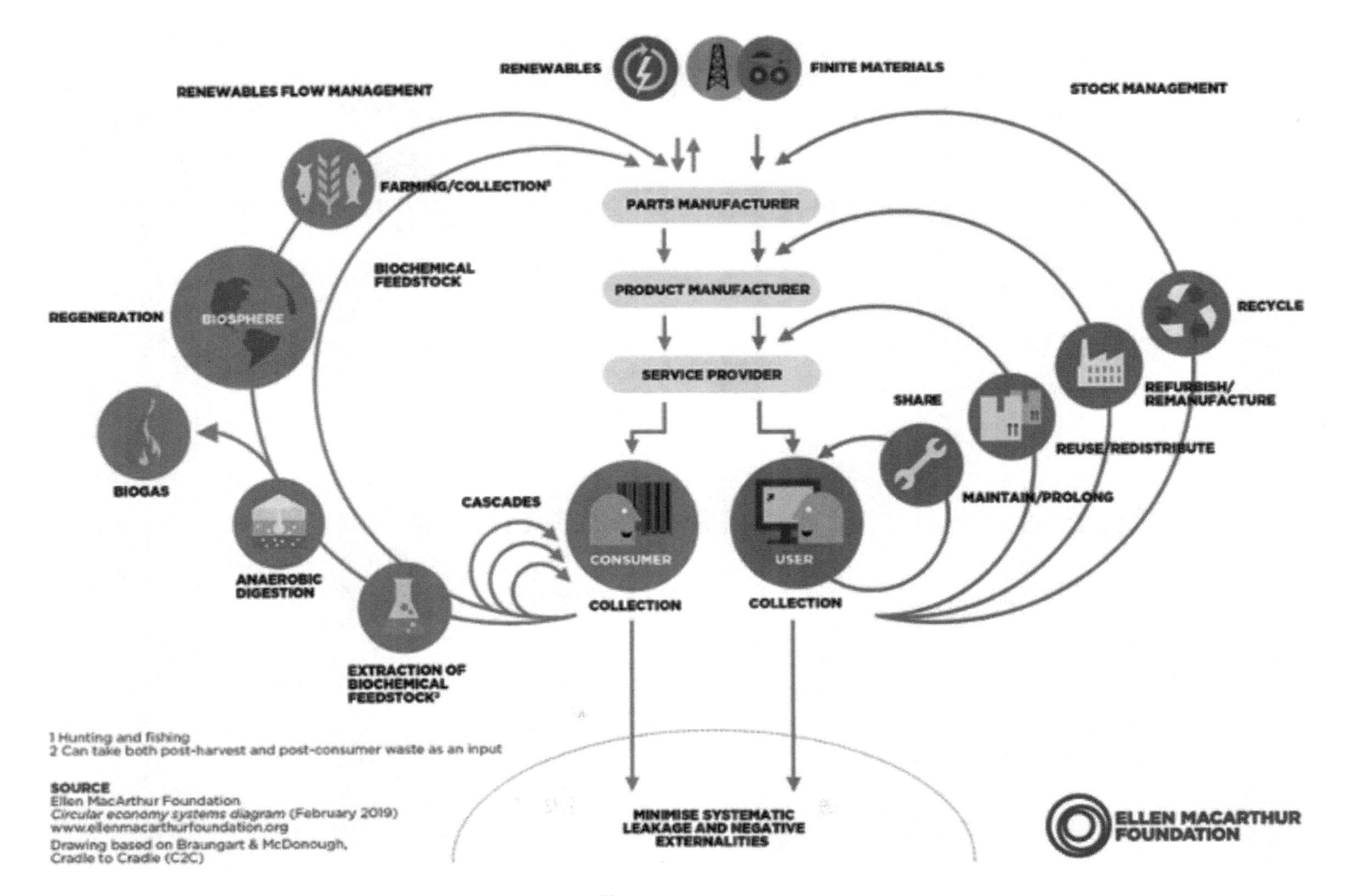


*Figure 3.8* The butterfly diagram: visualising the circular economy[8]

However, to make such a shift, the global economic system must change, starting substantially for each sector separately. Aviation and aerospace can follow that model, adjust and adapt accordingly. Material flow, resources used, type of energy consumed, life cycle impact, waste management and reusing or recycling of materials are some basic principles aviation can follow under a circular economy model.

Before establishing the application of a circular economy model to the aviation and aerospace industry, it is necessary first to review and understand the fundamental concepts of a circular economic model.[8, 9]

### *From Linear to Circular Model*

For any sector to become economically sustainable, it is necessary to review the relevant concepts and models that support that shift. This shift requires transforming from a linear economic model to a circular one. Then it is vital to see how this transformation can apply to the aviation sector. What is the difference between a linear and a circular model? What are the necessary changes to make the aviation industry operate under a circular economic model? And most importantly, how can these changes be applied?

#### *What Is a Linear Economic Model?*

A linear economy traditionally follows the 'take-make-dispose' step-by-step plan. The concept starts with collecting raw materials and then transforming them into products and goods that should be used until they are discarded as waste. Value is created in this economic system by producing and selling as many products as possible. The linear sequence uses raw materials to make a product and then puts it in landfills as waste. This economic model follows the sequence: take (raw material), make (products), use (consume) and dispose (of non-recyclable waste). This is the most common model used until now not only in the aviation sector, and it has been proven unsustainable for natural resources, consumption and the environment.

Figure 3.9 depicts the basic linear economic model: when goods are manufactured and produced, they often take only one direction, and the product ends in landfills as waste. In circular economies based on recycling, waste materials are reused.[9]

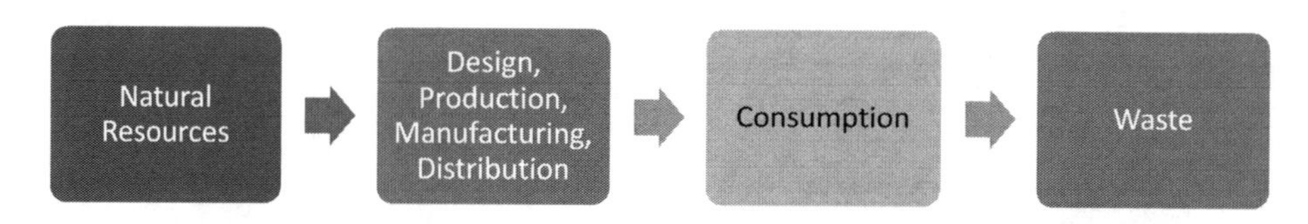


*Figure 3.9* A basic linear economic model

The linear model or the approach 'take-make-use-discard' model has been used for decades and is currently used to a large extent. However, waste treatment was a 'nonproductive' activity for product manufacturers, as money is invested without any economic returns. The concept of the linear economy is based on the 'cradle-to-grave' assessment. It reflects the impacts at each stage of a product's life cycle, from the time natural resources are extracted from the ground and processed through each subsequent stage of manufacturing, transportation, product use and disposal.[10] In the 1990s, this model was challenged by introducing waste minimisation, pollution prevention or cleaner production. It focused on reducing the waste produced rather than treating it when generated. It was a proactive (preventative) rather than a reactive approach. That change was undoubtedly one of the first steps towards sustainability. The concept of cleaner production initially focused on the factories, in production or process units, and the idea was to minimise energy resources and reduce waste emissions.

In creating and supporting economic sustainability for aviation business units, it is essential to look closely at the linear model that applies to the different types of aviation operations and start converting them to an economically sustainable model. The circular economy is the fundamental model to be used in commercial aviation, given its complex nature of interlinked and interrelated operations, from manufacturing to flights. The conversion from a linear to a circular economy requires first looking at the circular model and applying substantial changes to the aviation sector.

*The Steps to Circular Economy*

The circular economy aims to change the paradigm of the linear economy. It focuses on limiting the use of natural resources, limiting waste and increasing efficiency in the various stages of the product economy. The current scarcity of natural resources, the degradation of the natural environment, the expansion of the population and the necessity to cover its needs show the imperative shift to a circular economy.[11] The term circular economy is a new economic pattern for economic development, and it refers to a 'closed materials loop'. The value of circular economy relies on the sense of a continuous resource cycle, using wisely the available natural resources, considering environmental protection, social interference and economic development. The circular economy mandates that humans are part of the production cycle and that natural and energy resources are used considerably in that continuous loop to reduce the wasteful impact as much as possible. In that way, economic activities will have less impact on the natural environment and, consequently, humans.[12] Hence, in a broader context, the circular economy requires the social production and reproduction of processes and activities, always considering the natural environment and resources, saving and using and recycling wastes as much as possible. Clean production, material flow analysis, environmental and energy management, Life Cycle Assessment and other means are technical methods used for a circular economy, always aiming for social and economic benefits under the protection of the environment.[13]

*Figure 3.10* Recycle

The circular economy model is based on the 'reduce-reuse-recycle' principle.

- 'Reduce' refers to reducing energy input, use and resources during production stages and waste discharge reduction during production and consumption.
- 'Reuse' is about the prolonged utility time and service during the design and production process. It also refers to the reuse of waste. This can be done by simply reusing wastepaper to produce new paper.
- 'Recycle' means transforming waste into a new economic resource through chemical or physical processes and putting the new material into production. In a more industrialised or manufacturing environment, this can be done with scrap tires for producing another form of plastic. This is called industrial symbiosis.

This circular approach transforms the linear economic model into a new transformational and sustainable economy. Applying a circular economy model to the aviation industry requires a thorough analysis of the aviation sector's specific operational domains. It is necessary to identify the operational processes and how a circular model can be applied, considering the 'reduce-reuse-recycle' approach as a basis. Then the material flow analysis will show the step-by-step process to transform the whole aviation system.[13]

**What is Industrial Symbiosis?**

*Industrial symbiosis* is when wastes or by-products of an industry or industrial process become the raw materials for another. This concept allows materials to be used sustainably, to reduce waste and support economic circularity. The transition to this economic model can enhance market competitiveness, sustainability, resource efficiency and security while reducing greenhouse gas (GHG) emissions. Applying an industrial symbiosis model can create a more robust network than the conventional systems. The energy and materials cycle can go continually with minimal waste products. This process reduces the environmental footprint of the industries involved. Raw materials are less necessary, and waste disposal in landfills is reduced. It creates value for materials that would otherwise be discarded. Hence, the materials remain economically valuable and useful for a longer time than they would be in traditional industrial systems.[14]

***The Benefits of Industrial Symbiosis to the Aviation Industry***[14]

*Industrial symbiosis* is an approach that does not apply only to the environmental aspect of an organisation's sustainability. The concept of industrial symbiosis, especially for the aviation, requires establishing and managing close cooperation with governmental bodies and other industrial entities and even getting acceptance from the general public. Its purpose is to exhibit the social, economic and environmental benefits such an approach can achieve. It is necessary to demonstrate good waste management strategies that aim to change how the aviation system operates and ensure cost efficiency at all levels. However, the change must come from the 'inside'. That is why, apart from the waste management strategies and the external cooperation among different stakeholders, it is also necessary to achieve internal stakeholder engagement. Four primary benefits can arise from applying industrial symbiosis to the aviation.

**The impact from waste reduction**

- Reduce environmental waste through recovery, reuse and recycling
- Reduce impacts and risks from waste sent to landfill or remain untreated.

**Economic value**

- Use waste to create economic value. Reuse and recycling can reduce costs for the company, unless there are new ways to sell or trade these materials that are not used anymore with other industries and companies.

**Environment**

- Reduce emissions from air transportation and raw material extraction
- Reduce the use of conventional fuels and NOx, SO2, CO2 emissions
- Use of Sustainable Aviation Fuels in aircraft
- Aviation facilities, airports, hangars, airline offices and manufacturing areas can apply measures for energy efficiency, create renewable energy sources develop plans that will support electricity consumption in these facilities.

**Knowledge and skills**

- Increase knowledge and practical skills on waste management
- Promote a culture for sustainable growth in the aviation sector
- Create horizontal trainings regarding sustainability for different aviation organisations

The impact reduction concerns the environmental output of the aviation in the after-use activity or processes. It focuses on waste management and the reduction of any by-product disposed in landfills. For example, a maintenance facility can sell old aircraft tires to other industries and reuse them for several processes. Metals that are redundant for further application in aircraft systems can be sent to factories for melting and creating new materials. As in all systems, the same in aviation, the human element's role is critical. People are the triggering factor behind everything. The industry should train and educate its staff. Employees must be familiar with any innovative approaches – such as sustainability or industrial symbiosis – their company decides to pursue. When people know and understand, they can support any new initiatives within their system. In that way, any relevant action will be even more cost-effective.

The Air France–KLM Group adopted the economic circularity model in 2015. The group developed a strategy which acts on five primary areas:

- Redesign catering services to properly separate wastes.
- Reduce the food packaging
- Replace printed manuals with digital ones.
- Reuse seats and onboard entertainment systems in other systems or aircraft.
- Recycle reusable equipment, such as trays, drawers, blankets, trolleys, etc.

The PAMELA (Process for Advanced Management of End-of-Life Aircraft) was a project co-funded by the European Commission, and it presented a management approach launched by Airbus to handle end-of-life aircraft. The PAMELA project included three stages:

- decommissioning;
- disassembly; and
- smart and selective dismantling.

The results of the PAMELA project demonstrated that 85% of an Airbus A300 aircraft weight can be recycled, reused or recovered as secondary raw materials.

### What Can the Aviation Sector Do to Change?

The Ellen MacArthur Foundation summarises the three principles of the circular economy in a similar context as the 'reduce-reuse-recycle' model, further explained in Figure 3.8 with the butterfly diagram. Products' end of life must be redefined so their durability and usability are extended, reducing the environmental effects of new product manufacturing. In a technical environment, where materials, components and even mechanical systems are, the role of appropriate maintenance, refurbishment, reuse and recycle are ways to extend the life cycle of products. They will no longer be considered waste, rather than helpful resources or raw material input for manufacturing. This approach is precisely what aviation can adopt. The second circular economy principle relies on renewable or waste-derived resources and energy. The third principle is to build resilience through energy and material diversity. Supply chains can be redeveloped to reduce raw material resources and energy consumption, decreasing environmental effects. Some aviation manufacturing domains already apply the basic principles of the circular economy. This new approach could reform the whole aviation supply chain.

A good example is the use of 3D printing for manufacturing aircraft parts. In some cases, the materials to create lightweight structures can be 40–60% lighter, reducing fuel consumption and CO2 emissions. Such materials must have been approved by the appropriate authorities, like EASA, FAA or any other national authority. Inevitably, the circular economy is a new and prominent concept, expanding to the aviation sector and many other industries. Even though its application is still not popular and widespread, the circular economy model provides valuable opportunities for change. Despite the industry's COVID-19 hit in 2020, the sector is expected to return to its estimated growth levels, with an average annual growth rate of 4.4%. This is envisioned through the projection of delivered aircraft from large manufacturers. The increase in aircraft manufacturing at that level implies an enormous number of resources needed, with a simultaneous increase in energy consumption, emissions and waste. Hence, it is imperative to transition from a linear to a circular economic model. The aviation industry is a sector that can embrace the circular economic model from the conception of a new aircraft development to the point of its decommissioning. A cradle-to-cradle model for all stages of the life of an aircraft is necessary. However, that same model should be applied in all aviation operations, from the aircraft design and manufacturing processes to the end of its removal from flying.[15]

#### *Elements for a Sustainable Growth in the Aviation Industry*

Aviation industry does not only require economic growth that will come through profit. The change in the industry must rely on sustainable growth, considering the basic elements for industrial growth, but with respect and consideration of the three main pillars of sustainability.

- **Productivity**. Incentives for greater efficiency in the use of resources and natural assets, in aerospace manufacturing and aviation maintenance, including enhancing productivity. Waste reduction sensible energy consumption and use resources to their highest value while aviation manufacturing operations support productivity and sustainable growth, with cost efficiency.
- **Innovation**. Opportunities for innovation that allow for new ways of creating value in almost all aviation domains value chain and address environmental problems. Innovation can cover a broad spectrum of technologies, from aircraft, equipment, software to renewable energy technologies in facilities and smart electricity systems in offices.
- **New Markets**. Creation of new markets by stimulating demand for air services, aerospace manufactured products, aviation maintenance, etc., creating also new job opportunities.
- **Confidence**. Create and enhance confidence among external shareholders and internal stakeholders. Any sustainable action and resultant growth comes with confidence.
- **Stability and Resilience**. Aviation can operate in a sustainable way first for survival and then for resilience.

### Conclusion

Economic sustainability refers to the long-term development of an organisation or system in such a way that it does not affect the other two pillars, environment and society, or other relevant concepts thereof. Economic growth is a crucial element for the viability of any aviation business entity, but economic sustainability should always respect the other two pillars. Economic viability can inevitably support actions that can embrace the social aspect of sustainability, both internally and externally. Furthermore, environmental management practices, waste management or energy efficiency investments can only be supported through an economically viable system. Economic sustainability in aviation means thinking about economic development and profitably, supporting the activities and people of the industry, but at the same time, respecting the people, both internally and externally. This can happen while protecting and supporting the natural environment. The five capitals model explains various principles of sustainable development within the economic concept of wealth creation, or 'capital'. Aviation is a valuable industry in the global market. More than 30% of all trade is shipped by air, making air transport a critical business factor worldwide. There are several benefits of aviation to the world economy and the local and global communities. From trade to facilities to employment such as, pilots, technicians and ground support staff, etc. That is a very large population of employees that rely on the aviation sector. At the same time, air transportation supports local economies, people in need and remote societies and areas. It is an undoubtful contribution to humanity; however, the economic effects of the industry can expand beyond that, if economic circularity becomes part of aviation core business values. Aviation does not only need economic growth that will come exclusively through profit. The shift in the industry can rely on sustainable growth, considering the basic elements for industrial development and progress but with respect and consideration of the sustainability's three main pillars.

**Key Points to Remember**

- Economic sustainability refers to practices that support long-term economic growth, not only without negatively affecting the other two pillars of sustainability, environment and society but supporting them too.
- Aviation is a valuable industry in global economic development. Over 30% of all trade is sent by air, positioning aviation as a critical business factor worldwide. It supports national economies to become part of the global economy, offering direct benefits through its economic growth.
- The economic benefits of aviation are divided into three main categories: direct, indirect and induced benefits. These three benefits of aviation's activity impact economic growth through five main areas.
- The aviation industry is a broad sector that gives employment to many people across the globe. It supports 87.7 million jobs, either directly or indirectly, which are linked simultaneously with the direct, indirect or induced financial impacts.
- Different aviation professions constitute this number, including aviation operators, air operators' employees, such as pilots, cabin crew, ground services, catering staff, maintenance staff and all managerial and administrative staff, aerospace and manufacturing employees, air traffic controllers and executives.
- Circular economy is an interdisciplinary concept that involves notions from natural sciences, social sciences, technology and economic values. In fact, circular economy is a systemic approach to economic development designed to benefit businesses, society and the environment.
- Material flow, resources used, type of energy consumed, life cycle impact, waste management and reuse or recycle of materials are some of the basic tools that aviation can use under a circular economy model.
- The circular economy requires the social production and reproduction of activities, always considering the natural environment and resources saving and using, and the recycling of wastes as much as possible.
- Industrial symbiosis is the process by which wastes or by-products of an industry or industrial process become the raw materials for another. This concept allows materials to be used more sustainably and supports creating a circular economy. The transition to this economic model can enhance economic competitiveness, sustainability, resource efficiency and security, reducing greenhouse gas (GHG) emissions.

## Chapter Review Questions

3.1 Explain what economic sustainability for aviation and aerospace industry is.

3.2 Consider Table 3.1 and the seven checkpoints for maximising aviation growth through a sustainable manner, and align them with the Table 1.1 which presents the 17 SDGs. Explain how each one of the seven points supports a different sustainable development goal.

3.3 ATAG states in their 2020 Aviation Benefits Report that 'Tourism is fast becoming the world's largest industry, and air transport plays a vital role. Conservative estimates suggest that aviation supports 44.8 million jobs within tourism'. Explain the effect of aviation's economic sustainability to tourism.

*Figure 3.11* Airplane no. 3 pointing right

3.4 Identify some benefits of circular economy in the aviation and aerospace industry.

3.5 Give an example of industrial symbiosis relevant to the aviation or the aerospace industry.

3.6 Which UN sustainability goals can be met through the application of a circular economic model for the aviation sector?

3.7 Explain how economic sustainability through the circular economy can support the other two pillars of sustainability, environment and society in aviation.

3.8 Apply the 'reduce-reuse-recycle' model to one aviation operational domain.

3.9 Explain how the second principle of the circular economy, as the Ellen McArthur Foundation presents it, can support the environmental and social pillar of sustainability.

## Project Case Questions

3.10 In 2010 when the Eyjafjallajökull Volcano erupted, the World Bank (2010) stated that 'The impact on producers of flowers and fruit and vegetables in African countries such as Kenya, Zambia and Ghana were reported widely, with delays in transportation meaning large quantities of fast-perishing produce rotted'. Describe the role of aviation's economic sustainability to the local communities in these African countries. Explain extended effects to the population from the cease of air transportation due to the volcano eruption.

3.11 After you have reviewed the model of a circular economy and the main elements to create such a model, examine the graphic representation of the circular economy model from Ellen McArthur Foundation. Using your critique and knowledge about sustainability and considering the main structure and activities of the aviation, create a similar theoretical model, addressing to a specific aviation business activity or type of operation.

3.12 Make relevant research on how a Maintenance, Repair and Overhaul organisation operates. Then select a series of work tasks performed. Provide an example of industrial symbiosis between an MRO and another type of industry, not necessarily in the aviation sector.

## References

[1] Binns, J. (2020, August 21). *The five capitals – a model for sustainable development*. RRC International. Retrieved February 18, 2022, from https://blog.rrc.co.uk/2018/07/30/the-five-capitals-a-model-for-sustainable-development

[2] Uniting Aviation. (2018, July 11). *Aviation benefits: Contributing to global economic prosperity*. https://unitingaviation.com/news/economic-development/aviation-benefits-for-a-better-future/
[3] ICAO. (2019a). *Aviation benefits report*. www.icao.int/sustainability/Documents/AVIATION-BENEFITS-2019-web.pdf
[4] ATAG. (2005). *The economic and social benefits of air transportation*. ICAO. www.icao.int/meetings/wrdss2011/documents/jointworkshop2005/atag_socialbenefitsairtransport.pdf
[5] ATAG. (2020, September). *Powering global economic growth, employment, trade links for sustainable development despite global crisis*. Aviation: Benefits Beyond Borders. https://aviationbenefits.org/media/167517/aw-oct-final-atag_abbb-2020-publication-digital.pdf
[6] ATAG. (n.d.). *Economic growth*. Economic Growth: Aviation: Benefits beyond Borders. https://aviationbenefits.org/economic-growth/
[7] Talbot, F. (2021, October 20). *The impact of COVID-19 on the airport business and the path to recovery*. ACI World. https://aci.aero/2021/07/14/the-impact-of-covid-19-on-the-airport-business-and-the-path-to-recovery-2/
[8] Ellen McArthur Foundation. (2019, February). *The butterfly diagram: Visualising the circular economy* [Illustration]. Ellen McArthur Foundation. https://ellenmacarthurfoundation.org/circular-economy-diagram
[9] Liu, L., & Ramakrishna, S. (2021). *An introduction to circular economy* (1st ed.). Springer
[10] European Environment Agency. (n.d.). *cradle to grave*. www.eea.europa.eu/help/glossary/eea-glossary/cradle-to-grave
[11] Solar Impulse Foundation. (n.d.). *Circular economy*. https://solarimpulse.com/circular-economy-solutions
[12] McDonough, W., & Braungart, M. (2005). *Cradle to cradle-remaking the we take things. ACCA and China-U.S. Center for sustainable development*. Tongji University Press (In Chinese)
[13] Yifang, L., Renchen, R., & Jian, X. (2007). *The input-output analysis of the circular economy*. International Input-Output Association. www.iioa.org/conferences/16th/files/Papers/Liu-287.pdf
[14] European Commission. (2018). *Industrial symbiosis*. European Commission|Environment. https://ec.europa.eu/environment/europeangreencapital/wp-content/uploads/2018/05/Industrial_Symbiosis.pdf
[15] ICAO. (2019b). *Introduction to circular economy*. www.icao.int/environmental-protection/Documents/EnvironmentalReports/2019/ENVReport2019_pg275-278.pdf
[16] European Commission. (2007). *Life 3.0 – life project public page*. Life Project Public Page. https://webgate.ec.europa.eu/life/publicWebsite/index.cfm?fuseaction=search.dspPage&n_proj_id=2859

# 4 Aviation Social Sustainability

**Chapter Outcomes**

At the end of this chapter, you will be able to do the following:

- Explain aviation social sustainability.
- Identify the role of human and social capital in social sustainability.
- Determine the role of social pillar to aviation and aerospace sustainability.
- Identify Corporate Social Responsibility.
- Describe the link between aviation human factor elements with social sustainability.
- Apply Corporate Social Responsibility in aviation and aerospace industry.
- Describe what ESG means for the aviation sector.

**Introduction**

*Figure 4.1* Aviation safety

This chapter discusses the third pillar of sustainability and how the aviation and aerospace sector supports it. Social sustainability is the most challenging element to explain due to its broad aspect and the elements it can cover. The following sections explain how social sustainability can be met through aviation's best operational practices not only for society but also for those who are part of the aviation ecosystem. Understanding social sustainability is the starting point to comprehend how it influences the aviation sector. Safety is of paramount importance in our industry and inevitably plays a significant part of social sustainability since a sustainable aviation system must continuously operate under safety. As of the ICAO Safety Management System Manual, aviation safety is achieved through a well-organised and structured system, with qualified and responsible personnel that has all the essential resources available and meets all the regulatory safety requirements. Nevertheless, to be and remain safe, an aviation system must also be sustainable. In addition, a significant component of social sustainability is Corporate Social Responsibility (CSR). CSR within aviation is explained, analysing also the four types of CSR: environmental, ethical, philanthropic and economic responsibility. Trends such as '*flight-shame*' – *Flygskam* – movement is explained, along with the '*greenwashing*'. Lastly, the

DOI: 10.4324/9781003251231-4

newly introduced concept of environmental, social and governance (ESG) is presented as part of social sustainability and a link connecting all three pillars under the sustainability 'umbrella'.

## What Is Social Sustainability?

Social sustainability is the third of the three pillars. The concept of social sustainability refers to an entity's impact on the society or people. *Society* is the external environment in which a company acts and operates. How a company or sector operates always affects society either positively or negatively. A positive reaction would be to support the local society. This support comes with employing people from the local community, buying goods and resources from the area where the organisation is placed and protecting the local natural environment. Protecting the natural environment would mean having systems that will not pollute the water, soil or air and protect wildlife and natural resources. It is the part of sustainability that considers humans, society and their environment. Under this framework and the interaction between people, society and the environment, it is logical to consider the economic element. An organisation that considers the economic benefit only and hires employees only from remote areas, not the local society, leads to low economic activity for the local social system. If the local environment is not a priority for an organisation and acts as an unconscious polluter, the local society will suffer long-term effects. The natural environment will be destroyed; therefore, the natural capital will be highly impacted. It is often reported that when organisations and industries pollute a natural environment, the residents may suffer severe health effects from consuming goods produced in these polluted natural environments.

At the same time, social sustainability does not refer only to the external social environment of an organisation. It refers to the people that work and operate within that organisation. The conditions within a working environment may affect the people employed there. For example, the aviation maintenance activities require the use of paints, solvents or other various types of materials to fulfil the necessary industrial activities. In that case, employees must be protected. Their protection relies on the supply of equipment that will eliminate or minimise their exposure to these hazardous health conditions. Social sustainability relies on humans, their well-being, ethical treatment, equity and respect.

As we have seen in Figure 1.4 of Chapter 1, the social aspect of sustainability includes proper education, equality among and within communities, gender and race equality, social development and justice. As briefly explained earlier, social sustainability is interlinked and interrelated with the other two 'circles' of sustainability, environment and economy. To make the aviation sustainable, the sector must be able to provide fair and respectable conditions for humans, either internally or externally. However, it is not always obvious or clear how to make a complex system like the aviation sustainable. Hence, the interrelations and interactions with the other two pillars should be carefully identified in an effort to meet the fundamental sustainability principles.

### *Human and Social Capitals for Aviation*

In Chapter 3, we explained the five capitals. Those capitals are vital elements for a business to deliver its products and services. Relevant to the concepts addressed in this chapter, human and social capital are presented, with a comprehensive approach applied to the aviation and aerospace. As a reminding note, social capital refers to the availability of mechanisms that can support and enhance human development and relationships, enhance communication efficiency both in and out of a company and generate valuable networks. Such an approach can bring benefits that will strengthen the trust and values of the employees within the organisation. Human capital is also a social issue within a business environment. Human capital signifies that a company supports

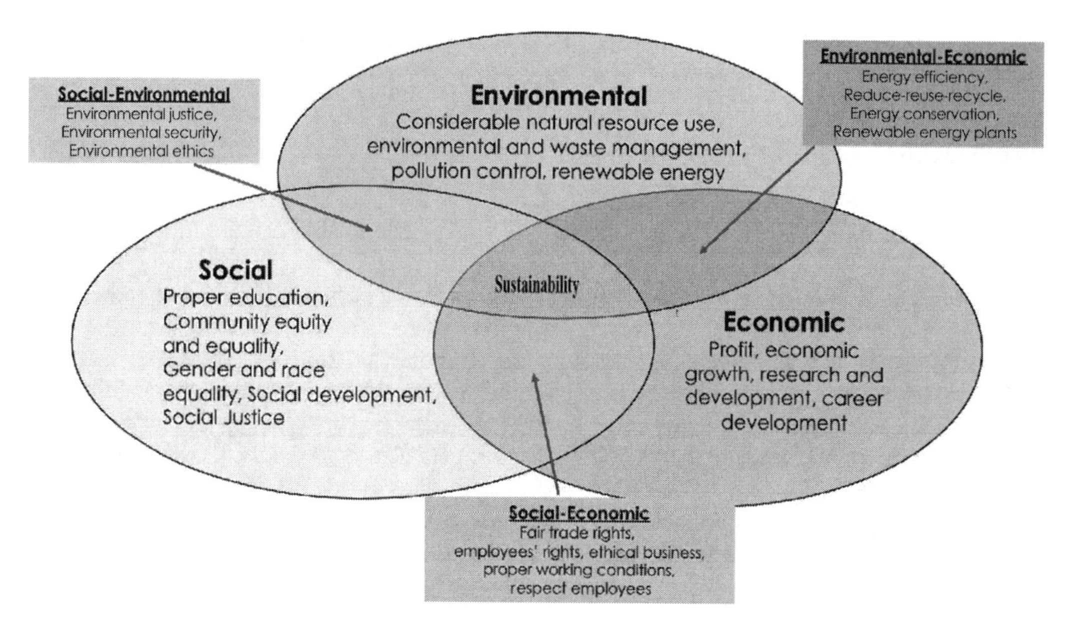


*Figure 1.4* The three circles of sustainability and their interrelations

employees' healthy working conditions and meets satisfactorily working standards. It requires a suitable trained and educated workforce to operate. The purpose of human capital in an organisation is to look after the employees. However, how important is it for aviation and aerospace to address the social and human capital models, and most importantly, do they support social sustainability for the industry and how? This is where safety enters and becomes part of aviation social sustainability.[1]

### *The Social Capital for Aviation*

The aviation and aerospace industry's social capital includes all these components that concern the development and preservation of the organisational culture. In human factors, the organisational culture is explained with the phrase 'The way we do things here'. It may sound – or read! – like a simple phrase or a very brief explanation for the term 'culture'; however, it hinders a lot more than that. The way 'we do things' in a company means how things are handled, how well problems are communicated, how comfortable employees feel communicating with their supervisors or other team members and how confident they are that their voice is heard within the system they work and valued – how they trust the management system and if the system trusts them.[2] The organisational culture in an aviation environment is a broad concept that eventually affects aviation safety. The way an organisation operates is mirrored in its culture. When this culture does not support social capital, meaning these mechanisms that can enhance communication and internal or even external networks, the social capital is not only undervalued but also in danger since safety is compromised. Environmental conditions in an aviation environment are often a human factor leading to an incident or an accident. Such an event will affect not only the humans involved but it also often results in financial liabilities due to fines or loss of property or, even worse, the loss of an aircraft. A chain of events can lead to an outcome which may initially affect the social aspect of the organisation – meaning employees, local society and even passengers – but eventually will affect

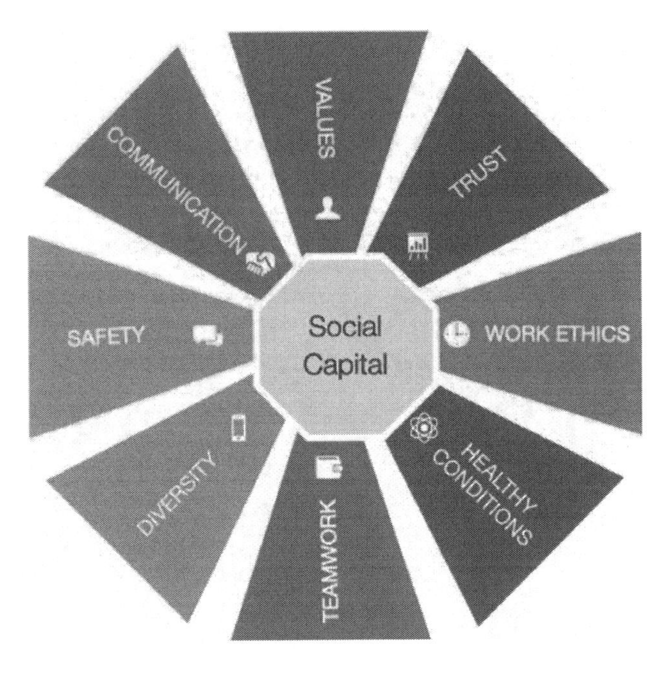


*Figure 4.2* Social capital

the organisation financially and, depending on the scale of that, its viability. The elements that form the organisational culture in an aviation system relies on teamwork, proper communication, healthy working conditions, ethics, trust and values and, most importantly, safety, for the personnel and for the users of the final product – the aircraft.

Therefore, all starts and ends with social capital, the values and trust of people to the company they work for. Do they feel they can trust their employer? Are they treated with respect? Do they work in healthy conditions with proper protective equipment? These questions signify what the company's culture is or should be. If the organisational culture supports proper behaviour and actions, social capital is highly valued supporting the company's social sustainability. If not, then sustainability is not one of the company's priorities, and consequently, there is plenty of space for change and improvements.

> The organisational approach to aviation safety was introduced during the mid-1990s when safety started to be seen from a systemic perspective. It began to consider organisational factors, not only personnel or technical factors. The term 'organisational accident' was also introduced. This perspective considered the impact of organisational culture and policies regarding the effectiveness of safety risk controls.

From the beginning of the twenty-first century, the safety approach changed in the aviation world. Up until then, safety was considering mainly individual's performance. However, due to the growth of the aviation industry, its maturity over the years and its complexity, the approach became more 'total', as we could say, or in different words, holistic. As mentioned in Chapter 1, the holistic approach means that the whole system is embraced, examined, changed and corrected if necessary. We used that same word to explain the purpose of sustainability as a holistic approach to a system. The holistic approach, either seen from the side of sustainability or the organisation side, has one priority – to look at the whole system, its processes, procedures, the way people work, the culture within that system and many more. Thus, humans are part of that holistic perspective, which is why they should be valued and feel trust in their workplace. We cannot talk about safety or organisational culture in aviation if humans, particularly employees, are not considered part of that system. Nobody can expect a high-safety performance if humans are not valued as they should. The same goes for sustainability. In a sustainable company, humans have a vital role in applying the new changes and supporting sustainable actions. If they are undervalued or exposed to unhealthy and unsafe working conditions, the system they work in is not sustainable nor safe.

*The Human Capital for Aviation and Aerospace*

Human capital mainly refers to factors that support the human element aiming to create wealth for the sector. Wealth is not necessarily a wrong term or a term that signifies something is happening as a burden to something or someone else. Wealth is something that all entities and companies pursue to stay viable, continue operating and, at the same time, employ people that are part of society. As a continuous loop, society is supported economically by the activities of that company. The wealth must not only focus on profit, meaning it should support the other two pillars – society and environment.

Hence, an economically viable company should provide its employees with the required working conditions and resources. The concept of human capital includes more practical factors such as necessary level of knowledge and skills for employees, opportunities for professional development and training, adequacy of trained staff and technical resources, such as tools, software or any equipment needed to do the job.

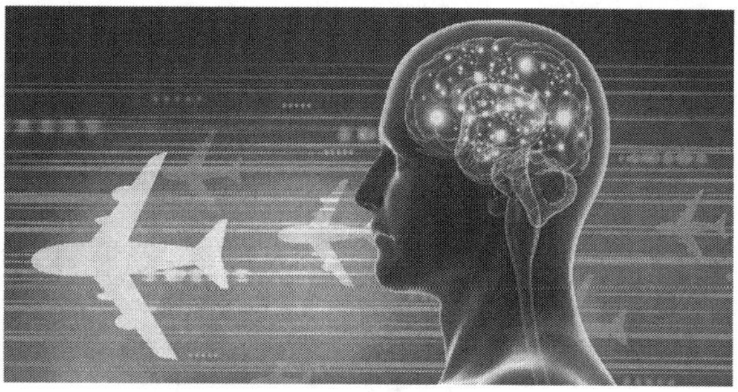

*Figure 4.3* Human factors in aviation

Taking these factors to the aviation environment, even by the regulation itself, or the theory of human factors and how to prevent mistakes and errors, it is apparent that they are necessary if not mandatory. For example, in an aviation maintenance environment, skilled workers are essential. Depending on their role, position and tasks performed, they must hold the relevant aviation maintenance license with the relevant working experience.

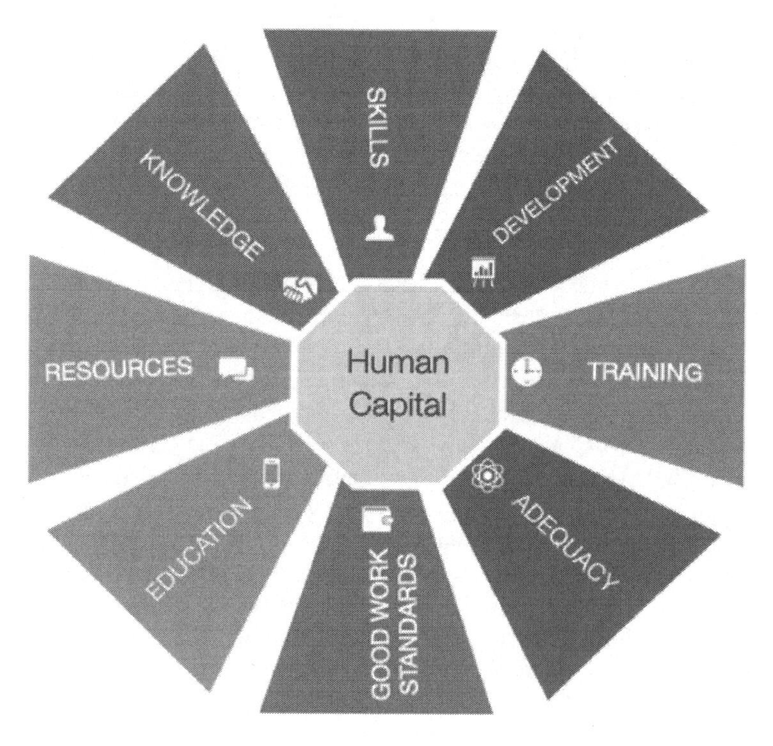


*Figure 4.4* Human capital factors

An aviation maintenance technician must stay current with the regulatory requirements. Under the EASA regulation, aviation maintenance technicians must pursue recurrent training for subjects, such as aviation legislation, human factors and other topics, every two years to keep their licenses active. If this training cannot be fulfilled internally, the maintenance organisation can offer that through a Part 147 maintenance training organisation and cover the related expenses. Similarly, as technology also advancing fast in the aviation maintenance domain, employees involved in maintenance tasks should be able to follow relevant technological advancements. They should have access to relevant training and the technology compatible with their working tasks and their responsibilities. All these examples are elements of the human capital investments necessary in the aviation maintenance industry when pursuing a sustainable pathway. Both social and human capitals refer to the social issues that an organisation should address in its efforts to act sustainably. Inevitably, the proper support of the social element will bring economic benefits to any company's development.

Aviation social sustainability encompasses employment, tourism activity, cultural heritage acquaintance and connection of remote areas. Aviation as a sector can significantly impact society and the people employed in it. The social aspect of aviation sustainability refers to the benefits that reach the people working in and for aviation and those using the aviation industry services. Those services include the connectivity for tourism and other recreational purposes or business, including services to remote areas. Aviation social sustainability involves the offering of proper working conditions, support, resources and training to personnel while adhering to aviation regulations, with safety being the ultimate and highest goal. The environment must be protected from the broader aspect of aviation activities to reduce carbon footprint and allow society to enjoy the benefits of a clean natural ecosystem.

*Employment and Aviation Human Factors*

Aviation significantly impact the economy on a global scale. At the same time, the social impacts of aviation are equally important. Employment is one of the main economic drivers, especially for a local economy. The starting point is where cash will flow in a small town or village. The shops will start work and make a profit for their owners and so on. However, we should remember the importance of employment for people. Employment is about more than making money and covering needs, which is very important – yet not the only objective. Employment motivates people to get educated, study and become professionals and continue to study and remain current on the advancements and updates of various professions. The aviation industry offers employment to a great range of professions. As we have seen in Chapter 3, direct and indirect jobs are created inside and outside the aviation industry. Employment supports social sustainability through direct, indirect and induced economic benefits. Consequently, economic benefits must support society. That is the point of having a sustainable economy and a sustainable society.

Going deeper into the aviation environment, social sustainability, as part of a sustainable system, must have the appropriate conditions to protect its 'internal' social element – its employees. The aviation sector cannot be socially sustainable if it only considers its impact on the social environment it belongs. It must first and foremost consider the working environment and conditions of its 'internal' society. Hence, what are the elements that make a working environment a sustainable one for its employees? For aviation, sustainability is strongly linked with one of the basic principles the industry mandates – human factors. They are the factors that affect people while they perform their work tasks. The purpose of human factors is to improve the interaction between people and systems. The Dirty Dozen list refers to twelve of the most common human error preconditions or conditions that can act as precursors to accidents or incidents. These twelve elements influence people to make mistakes. A sustainable system will consider the risks associated with these errors and provide the necessary resources to mitigate these risks and provide safe aviation operations.

'Human factors' studies the human capabilities and limitations in the workplace. Human factors researchers study system performance, which includes the interaction of maintenance personnel, equipment used, written and verbal procedures and rules they follow and the environmental conditions where employees are placed. Human factors aim to optimise the relationship between maintenance personnel and systems to improve safety, efficiency and well-being.[3]

*Table 4.1* The Dirty Dozen[2]

| *Lack of Communication* | *Complacency* | *Lack of Knowledge* | *Distraction* | *Lack of Teamwork* | *Fatigue* |
|---|---|---|---|---|---|
| Lack of resources | Pressure | Lack of assertiveness | Stress | Lack of awareness | Norms |

Going more profound in analysing human factors and their role in social sustainability for employees' environment, let us look at the conditions vital for a sustainable working environment. The theory of human factors does not include only the following. There are more. However, their analysis here aims to show how they can affect or further support a sustainable working environment for aviation employees.

*Human Factors Affecting Aviation Social Sustainability*

Social sustainability must be supported internally, in an organisation's environment, creating the necessary conditions that will be respectful and ethical, striving to improve employees' work. Social psychology is one of the most critical factors affecting human performance in an aviation environment.

*Figure 4.5* The human element

This element is analysed from the perspective of how important it is to create a sustainable working environment for the aviation workforce. Social psychology comprises components that affect not only human performance but also the mental well-being of employees. Social sustainability requires the necessary conditions for employees to work in a decent and safe environment, respecting mental health. Every aviation organisation or company will have different 'ways of doing things', which is called organisational culture. The company will hold its philosophy, policies, procedures, selection and training criteria, quality assurance methods and culture. The impact of the organisation may be positive or negative. The aviation industry can encourage their employees (both financially and with career incentives) and take notice of problems that their staff encounter, attempting to learn from these and make changes where necessary or possible. On the negative side, an organisation from example may pressure its engineers to get work done within tight timescales and budgets or add extra flight hours to crew personnel without respecting crew rest schedules. A possible scenario for employees is to feel that this approach will conflict with their ability to sustain the quality of their work. These organisational stresses may lead to problems of poor relations, high turnover of staff, increased absenteeism and, most notably, more incidents and accidents due to human error in the system. If these stressors are not appropriately handled, mental well-being and physical tiredness will compromise their work performance, the company's performance and safety. Under these or similar conditions, employees are not working in a sustainable environment.

Factors that affect work performance and can increase or decrease productivity may have multiple aspects. However, having an organisational culture that tolerates harmful work practices will demonstrate how sustainable this working environment is and how committed the leadership is to creating and supporting a sustainable working environment. Harmful work practices include fitness and health stress, time pressures, extreme workload, fatigue and employees' inconsiderable use of medication, alcohol and drugs without the appropriate cautionary measures. In critical working positions, aviation personnel must be able to judge if their mental and physical status may or may not compromise their daily work. In addition, the company must be able to provide the appropriate working conditions that will not affect their mental or physical status. Working in a stressful company where employees are not respected or are not being ethically treated is not a sustainable working environment. Aviation personnel can experience stress for two reasons: the task or job they are undertaking at that moment or the general organisational conditions. Stress can be felt when carrying out specific tasks that are challenging. This stress increases by lack of guidance or time pressures to complete the job but reduces through careful management, supervision and training and even hiring

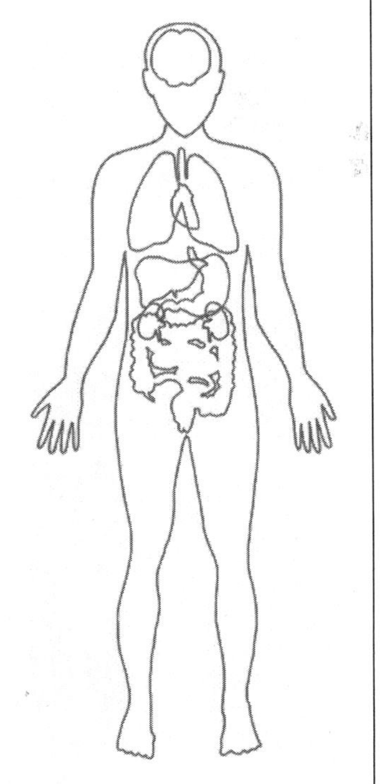

*Figure 4.6* The human body

new personnel with a specialisation at need. Suppose the company does not provide the appropriate resources, does not support employees to complete their work on time, correctly and safely, and is ignorant of stress and physical fatigue, this working environment is not a sustainable one.

A physical aviation environment and the conditions entailed include noise, fumes, illumination, climate and temperature, motion and vibration, the need to work in confined spaces and issues associated with the general conditions. Those examples are evident especially in a technical area work such as in or out of a hangar or an aircraft structure area. All these conditions may compromise work performance, and the person's physical status can be mitigated by protecting employees with the appropriate personal protective equipment. Earmuffs can reduce the sound intensity of loud noises. Protective goggles can protect the eyes from substances that can compromise eyesight in the short or long term. In addition, breathing apparatus or any other specialised equipment must be provided to employees when needed for their work tasks. The leadership of an organisation should be able to provide the resources to mitigate the effects of these conditions. Only then the working environment is sustainable. Despite the apparent direct consciousness that harmful conditions might have to an aviation worker, short or long-term health problems will cause work absenteeism or even work redundancy, lack of trust and rush to finish the work in progress. In that last option, it is possible for workers to 'cut corners' in their tasks, and this can even compromise a whole aircraft's safety. From an ethical point, it is also unacceptable for top management to neglect health-threatening conditions, creating an unsustainable work environment.

Another significant factor that affects work and employees' performance is communication. Communication in aviation is a critical element. It could be argued that there are three main functional areas where communication has a strong place. The first is the operations area where pilots, air traffic controllers and airport staff are all involved with the operational side of aviation, which is the flight itself. Efficient communication between people involved in the air operations is synonymous with safety.

*Figure 4.7* People's communication

Similarly, the second functional area in which aviation communication is critical falls under the 'umbrella' of human factors as well. For instance, aviation maintenance and maintenance management, crew resource management and ground operations also consider communication a vital element for safety. Safety management systems and human factors consider communication as an element that can enhance or decrease safety standards. In both SMS and HF, the interested parties include aircrew personnel, ground support staff, maintenance staff, production organisation, the policy authority or any other national or governmental entity related directly or indirectly with the aviation standards. For instance, the distribution of critical information of an Airworthiness Directive (AD) from the aviation authority (FAA, EASA, CASA, CAA, etc.) or a Service Bulletin from the aircraft type design holder to the respective operators – or in worst case scenarios, when an air crash accident occurs. Who is responsible to communicate what, to whom and how? From a sustainable standpoint, with lack of communication or not appropriate distribution of information, even a simple task may quickly disintegrate to include other human errors such as complacency, lack of knowledge, lack of teamwork, pressure, lack of assertiveness, stress, lack of awareness and norms.

## Aviation and Aerospace Corporate Social Responsibility

Corporate Social Responsibility (CSR) is part of social sustainability. As described earlier in this Chapter, Social Sustainability involves an entity's social impact, internally and externally. Corporate Social Responsibility (CSR) is part of the external impact of social sustainability. The most simplified definition of CSR is how a company benefits the external social environment, local or global. Corporate Social Responsibility activities can benefit the external social environment but the organisation as well. Positive actions can promote a company's relationships with the society through practical strategies that will boost environmental, social and economic outcomes. In addition, a correct CSR strategy with proven community benefits and can boost the company's reputation, offering in certain cases even some economic benefits.

Corporate Social Responsibility is part of the broader concept of social sustainability. Social sustainability for aviation includes the jobs and working conditions it provides to:

- its employees,
- its ability to connect people to remote global areas,
- the humanitarian use of aviation concerning air ambulance or disaster relief.

CSR often includes nonprofit activities that have a visible and well recognised benefit for the local or extended social element, as well. For example, many aviation companies establish nonprofit events to benefit and support the society or part of it. An event to rebuild a school in an area damaged by a physical catastrophe, an event to collect funds for charity or reforestation of areas that have been burnt from wildfires are some cases of CSR actions from aerospace manufacturers or airlines from time to time.

The broader concept of Corporate Social Responsibility is the operation of a business to create long-term value and support society while at the same time adhering to the ethical legal and public standards held by the larger society. Such business operations will aim to safeguarding the right and welfare of employees, offering safe products of reasonable quality, protecting the environment, avoiding bribery and corruption, and contributing to sustainable development. CSR extend to nonprofit support through philanthropic fundraising, or group actions, aiming to benefit the society or groups of society that are in need.

Corporate Social Responsibility (CSR) is a strategic business tool that can boost sustainability in the aviation sector. CSR is examined through different approaches, no matter the type of industrial sector. The four types are Environmental Responsibility, Ethical Responsibility, Philanthropic Responsibility and Economic Responsibility. While the definition of CSR is already explained in this section, and its applicability extends to the aviation, it is worthwhile to explain the four types of CSR from that aviation perspective. Many aviation companies, mainly airlines, embrace Corporate Social Responsibility due to moral convictions and often counterbalance the recent trend of *'flight-shame'*. Doing so can bring several benefits. For example, Corporate Social Responsibility initiatives can be a powerful marketing tool, helping an aviation company to position itself favourably in the eyes of consumers, investors, and regulators. However, in any case, companies must be honest about their actual efforts and results to be ethical, environmentally concerned and sustainably oriented. There are many cases in the news where a few airlines have not been honest and did not provide accurate information to the public on their environmental sustainability actions. Hence they were accused of *greenwashing* since their main aim was to fabricate a nice marketing image with inaccurate data. CSR initiatives can also improve employee engagement and satisfaction since, in particular, employee satisfaction is a major retention criterion for any company. Such initiatives can make a difference between airlines to attract investors or customers who hold strong beliefs that match those of the organisation. Additionally, Corporate Social Responsibility initiatives force business leaders to examine practices related to hiring and management processes, sourcing materials, and assessing the value of the product that is delivered to customers. This contemplation often leads to innovative and groundbreaking solutions that help a company act more socially responsibly and increase profits. Moreover, in a different concept than airlines, via CRS practices aircraft manufacturers may apply less energy-consuming processes with less waste production processes.

*'Flight-shame' or Flygskam*, comes from an anti-flying social movement created in Sweden in 2018. The movement's aim is to reduce carbon footprint and emissions from flying. The purpose of the movement is to feel responsible for your own carbon footprint and change travellers' behaviour, choosing a slower way of transportation.

Greenwashing is a form of advertising, claiming environmental practices and operations aiming to reduce environmental footprint without substantiating proof of what they do. The companies accused of greenwashing advertised that they were environmentally conscious only for marketing purposes, without actually making any effort in that direction, deceiving customers and society.

Environmental Responsibility

- Environmental responsibility refers to the belief that aviation organisations should behave as environmentally friendly as possible. It is one of the most common forms of corporate social responsibility. The term 'environmental stewardship' is often used to refer to such initiatives.
- Aviation, like all other industries, seeks to embrace environmental responsibility in several ways:
- Reducing pollution and flight emissions through the application of emission schemes (CORSIA, EU-ETS), use of single-use plastics in flights, water consumption in airports, and waste either from flights through reduce-reuse-recycle approaches or on a grander scale to provide proper aircraft decommissioning practices.
- Increasing reliance on RES, sustainable resources, and recycled or partially recycled materials in airports, airline offices, maintenance hangars, or aircraft manufacturing facilities.
- Offsetting negative environmental impact by planting trees, funding research and donating to related causes.

Ethical Responsibility

- Ethical responsibility ensures that an aviation organisation operates well and ethically, considering the necessary standards. An aviation organisation that embraces ethical responsibility aim to treat all stakeholders equally, with respect and provision of the safety standards, to respect their being, offering the products it claims it will, from employees to society.
- Respect for working conditions and being considerate of the impact its operations might have on lives is also part of ethical responsibility.
- Just culture, inclusion and respect are also included in the concept of ethical responsibility.

Philanthropic Responsibility

- Philanthropic responsibility refers to an aviation company's aim to actively make the world and society a better place.
- In addition to acting as ethically and environmentally friendly as possible, organisations driven by philanthropic responsibility often dedicate a portion of their earnings.
- Many aviation firms have donated to charities and nonprofits that align with their guiding missions. Others have donated to worthy causes that don't directly relate to their business.

Economic Responsibility

- Economic responsibility is the practice of a aviation company supporting all of its financial decisions in its commitment to do good in the areas listed above.
- The end goal is not to simply maximise profits but positively impact the environment, people and society.

*Figure 4.8* The different types of corporate responsibility

## Environmental, Social and Governance

Environmental, social and governance (ESG) is a green investment tool containing all the criteria and standards for a company's operations. Very recently, ESG has also made a significant appearance in the aviation. Investors with sustainability interests use ESG as a tool to screen potential investments.

Environmental criteria consider the actions taken or about to be taken towards environmental sustainability and carbon footprint reduction. The 'E' element in ESG also looks at a business's impact on the environment from resource consumption and operations, like carbon footprint and wastewater discharge, and any other activities that impact the environment.

Social criteria examine how a company manages relationships with employees, suppliers, customers and communities. Ethical working conditions, professional development support and resources are some elements considered in the 'S' part of ESG. The 'S' of ESG looks at how a business interacts with the communities. It also looks at internal policies related to labour, diversity and inclusion, among others.

Governance deals with a company's leadership, payments, regulatory compliance, audits, internal controls and shareholder rights. The 'G' of ESG relates to internal practices and policies that lead to effective decision-making and legal compliance.

ESG facilitates the long-term growth, aiming to attract talent, reduce costs and create trust with consumers.

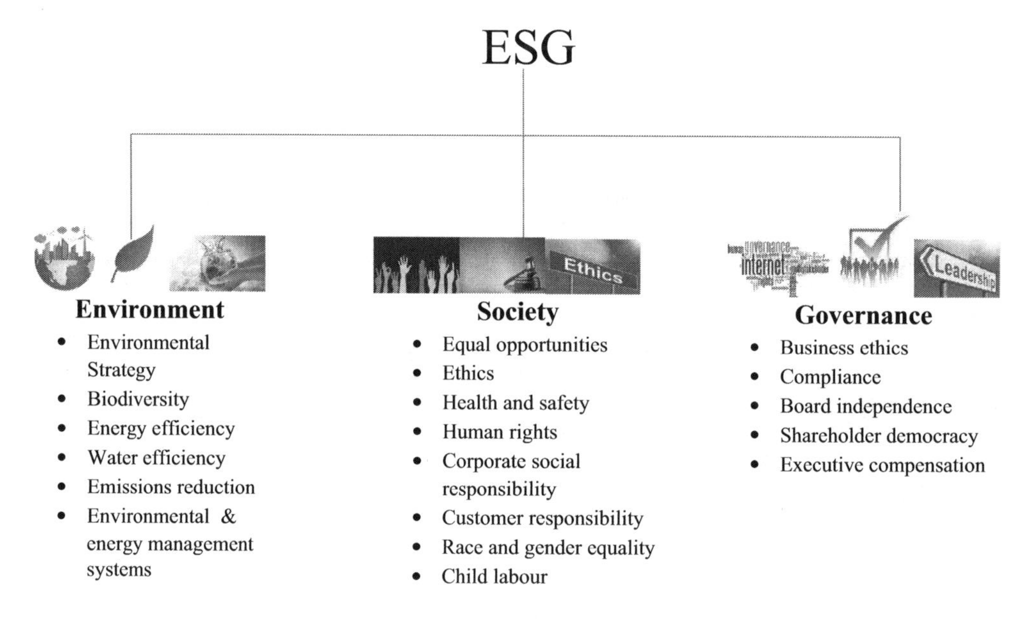


*Figure 4.9* The ESG model

## Conclusion

This chapter examined the aviation social sustainability and its application in aviation. It explained how social sustainability can be met through aviation and aerospace operations not only for society but also for those executing and who are part of aviation. Aviation safety was also explained and aligned as part of social sustainability. A sustainable aviation system should continuously operate under safety. Considering the aviation human factors theory and the ICAO Safety Management System Manual, aviation safety is aligned as part of aviation social sustainability. Social sustainability can be achieved through a well-organised and structured system, with qualified and responsible personnel that has all the essential resources available and meets all the regulatory safety requirements. Another important element of social sustainability is Corporate Social Responsibility (CSR). CSR was presented as part of aviation and aerospace companies' operations, and the four types of CSR – environmental, ethical, philanthropic and economic responsibility – were also analysed. Trends such as *'flight-shame'* – *Flygskam* – movement and the concept of *'greenwashing'* were explained. Lastly, the newly introduced concept of environmental, social and governance (ESG) was explained as part of social sustainability and a link connecting all three pillars under the sustainability 'umbrella'.

**Key Points to Remember**

- The social sustainability is possibly the most difficult aspect to explain due to its broad aspect and elements it can cover.
- The concept of social sustainability refers to the impact an organisation or entity has to the society or people. Society is the external environment, in which an organisation acts and operates.
- The way an aviation entity contacts its business can always affect the society positively or negatively.
- When an organisation considers the economic benefit only and hires employees that are from remote areas only and not the local society, this leads to a low economic activity for the local society.
- Social sustainability does not refer only to the external social environment of an organisation. It refers to the people that work and operate within that organisation.
- To make the aviation industry a sustainable one, it must provide fair and respectable conditions for the humans. It is not always clear how to make a complex system like aviation a sustainable one.
- The social capital for aviation includes all these components that are necessary for the organisational culture. In human factors, the organisational culture is simply explained with the phrase 'The way we do things here'.
- The human capital refers to factors that support the human element of a company, aiming to mainly create wealth for the organisation. The term wealth is not necessarily a bad term or a term that signifies that something is happening as a burden to something else. Wealth is something that all entities and organisations need to stay viable and operate now and into the future. Therefore, an organisation needs wealth to survive, to continue to function and to employ people that are part of a society, which is also supported economically by the activities of that organisation.
- The social sustainability aspect of aviation encompasses employment, tourism activity, cultural heritage acquaintance and connection of remote areas, among others. The aviation brings a significant impact to society and to people employed at the industry.
- The social aspect of aviation sustainability refers to the benefits that are reaching to the people that they are working in an aviation environment and to the people that are using the aviation industry services either for tourism or other recreational purposes or for business and necessity to provide support to remote areas.
- Aviation supports tourism as it transports people in closer or very distant areas, from people's origins. Social sustainability has as its core element, the society and humans. An industry that can support peoples' recreational activities, apart from business trips, offers a lot more than 'just a trip'.
- Corporate Social Responsibility is the operation of a business so as to generate profit for its owners and shareholders and create long-term value while at the same time adhering to the ethical legal and public standards held by the larger society. Those standards might be safeguarding the right and welfare of employees, offering safe products of reasonable quality, protecting the environment, avoiding bribery and corruption and contributing to sustainable development.
- Environmental, social and governance (ESG) criteria are a set of standards for a company's operations that socially conscious investors use to screen potential investments. Environmental

criteria consider how a company performs as a steward of nature. Social criteria manage relationships with employees, suppliers, customers and communities. Governance criteria explain the regulatory commitment, ethics and values a company must entail in its core business.

### Case Study: New MRO Facility in Kochi Area, India[4]

Air Works is an aviation maintenance company that in 2020 started a series of processes to get an approval to build a new Maintenance, Repair and Overhaul (MRO) facility in the Kochi Area in India from the Indian Aviation Regulatory Authority and EASA. Among the company's plans was to increase the capacity of its existing MRO unit at Hosur in Tamil Nadu from four bays to five bays within the next sixty days. Up until that time, the hangar could handle four aircraft simultaneously at Hosur with four bays. The MRO serves aircraft in the size of A320 or B737, including also private jets. The Indian carriers have a fleet of around 700 aircraft, which includes narrow-body aircraft like B737 and wide-body aircraft like B787, among them. Apart from facilities in Hosur and Kochi, Air Works has DGCA-approved facility at Mumbai for smaller aircraft (general aviation aircraft). The Kochi MRO will have two bays, but since they are quite big, they will be able to handle two private jets, two helicopters and two narrow-body aircraft simultaneously. The turnaround time – time taken to repair or overhaul – of an aircraft depends on what kind of checks one is doing. Lower-level checks that are termed as C1, C2 and C3 can be done in three to six days. More intensive checks that are termed in between C6 and C12 take about four to six weeks. The leasing period of an aircraft in India can be anywhere between three to eight years. The aircraft has to be in fit shape when it is delivered to the lessor after the use. Kochi is an enchanting city situated on the southwest coast of India in the state of Kerala. It is also a flourishing port city showcasing a rich blend of mesmerising natural beauty and vibrant culture. Considering the staffing requirements of an MRO, the maintenance needs to cover Indian fleets and all the topics discussed in this chapter, review and answer the case study questions at the end of this chapter.

*Table 4.2* Acronym rundown

| | |
|---|---|
| CSR | Corporate Social Responsibility |
| ESG | Environmental, Social and Governance |
| MRO | Maintenance, Repair and Overhaul |
| DGCA | Directorate General for Civil Aviation |
| EASA | European Aviation Safety Agency |

### Chapter Review Questions

4.1 See Figure 1.4 and explain what each intersection means (i.e. social-environmental and social-economic) for the aviation regarding the social component.

4.2 Provide a brief definition for social and human capital for the aviation industry.

4.3 Explain the role of organisational culture to social sustainability.

4.4 How do healthy working conditions link to social sustainability?

4.5 Briefly provide the similarities between the total system approach and sustainability.

*Figure 4.10* Airplane no. 4 pointing right

4.6 Analyse how environmental working conditions affect the social sustainability in aviation.
4.7 Provide an example on how the social capital can be part to social sustainability for airline staff.
4.8 Provide an example on how the human capital can be part to social sustainability for aerospace manufacturing personnel.
4.9 How does social psychology affect the other two pillars of sustainability?
4.10 Explain how social sustainability in aviation is related to the aviation safety for an aviation/aerospace entity.
4.11 Identify three factors that may impact work performance in an aviation environment, and explain how they affect a company's social element, employees and leadership.
4.12 What is CSR for an aviation/aerospace company?
4.13 Explain the four types of CSR and their role to aviation sustainability.
4.14 What is the ESG model?
4.15 What is the purpose of implementing an ESG investment tool?

***Case Study Questions***

4.16 Explain how the creation of the new MRO facility in Kochi supports social sustainability, justifying your response.
4.17 What are some benefits from this new MRO facility to the local society?
4.18 How can the development of the new facility support sustainability for the whole organisation?
4.19 How should the organisation operate to cover the social and human capital requirements?
4.20 What are some basic elements that the organisational culture should hold so it can support social sustainability in the new MRO facility?

## References

[1] IEMA. (2017, November 9). *An introduction to capitals*. Retrieved January 18, 2023, from www.iema.net/myiema/login?redirect=resources/watch-again/2017/11/09/an-introduction-to-capitals-relationship-capital
[2] Civil Aviation Authority. (2002). An introduction to aircraft maintenance engineering human factors for JAR 66. In *CAA*. Retrieved January 18, 2023, from https://publicapps.caa.co.uk/docs/33/CAP715.PDF
[3] ICAO. (2013). Safety management manual (SMM). In *ICAO*. Retrieved March 3, 2022, from www.icao.int/SAM/Documents/2017-SSP-GUY/Doc%209859%20SMM%20Third%20edition%20en.pdf
[4] Economic Times. (2021, February 6). Air works expecting regulatory approvals within 4–6 weeks for new MRO unit in Kochi: CEO. *The Economic Times*. Retrieved January 18, 2023, from https://economictimes.indiatimes.com/industry/transportation/airlines-/-aviation/air-works-expecting-regulatory-approvals-within-4-6-weeks-for-new-mro-unit-in-kochi-ceo/articleshow/80723288.cms
[5] The Five Capitals – a Framework for Sustainability. (2018, April 5). *Forum for the future*. www.forumforthefuture.org/the-five-capitals
[6] Cleveland, C. J., & Morris, C. G. (2015). *Dictionary of energy* (2nd ed.). Elsevier. https://online.hbs.edu/blog/post/types-of-corporate-social-responsibility
[7] What Is Environmental, Social, and Governance (ESG) Investing? (2022, September 27). *Investopedia*. www.investopedia.com/terms/e/environmental-social-and-governance-esg-criteria.asp

# 5 Sustainable Aviation Designs

## Chapter Outcomes

At the end of this chapter, you will be able to do the following:

- Explain the sustainable design of an aircraft and its phases.
- Explain what Life Cycle Assessment (LCA) for aviation is.
- Understand the use of LCA in the aviation and aerospace industry.
- Delineate the circular economy with aviation sustainable design and fuels.
- Understand the role of ISO 14044 in LCA.

## Introduction

The aviation industry has always been keen on new and innovative designs pursuing aircraft that can fly faster, higher and longer distances. That notion is evident throughout aviation history, especially after the Second World War. At that time, commercial aviation proliferated, mainly using ex-military aircraft designs to transport people and cargo. Even though they have been developed as 'instruments of war', they served well in the economic boom of the post-war era. However, their conceptions and design had not taken seriously into account the environmental impact. Flying faster means more powerful engines, which produce more noise and need more fuel. That was a reasonable assumption and an accepted design compromise, given the limitations of the available technologies of this era. Nevertheless, as the aviation industry matured over the years, new emerging needs surfaced, and a new approach was necessary to balance design requirements in a sustainable outcome. Sustainability comes into place as a broader context to address those needs at the design stage of a product and is critical for all aviation parts, systems and end products. While new propulsion systems, more efficient engines and improved aerodynamic performance offer significant reductions in CO2 emissions, a sustainable product also requires

*Figure 5.1* Airplane

DOI: 10.4324/9781003251231-5

optimising existing processes via methods like Life Cycle Assessment (LCA). What is LCA, and what are the different phases and applicability? An aircraft's life cycle can be examined through three phases: the aircraft design and production, the aircraft use and the aircraft's end of life or decommissioning. Thus, the sustainability concept embraces all aspects and the life cycle of the product from design and conception to end of life and decommissioning.

## Aircraft Designs

Improvements in aerodynamic propulsion and structures have a direct link to aircraft emissions reduction. Improvements in systems design and manufacturing technology are also the key to achieve future aircraft CO2-reduction goals. At the development stage we are now, aviation aims to adopt technologies and designs, aerodynamic structures and new fuels for reduced emissions and fuel efficiency.

### *Aerodynamics*

In the past years, additional advanced long-range twin-aisle airplanes with significant improvements in each area have entered operational service – the Boeing 787-9 and -10 and the Airbus A350-900 and -1000 – while the new Boeing B777-9 aircraft with a completely new composite wing is being prepared for certification testing. Moreover, several recently introduced new single-aisle aircraft (such as the Airbus A220-100 and -300) and several derivative aircraft with primary propulsion and airframe technology upgrades (such as the Airbus A320neo and A330neo, the Boeing B737MAX family and the Embraer E-Jets E2) have entered operational service and provide substantial reductions in fuel consumption. Progress is being made in developing and testing practical aerodynamic and manufacturing technologies, enabling reduced skin friction through laminar and conditioned turbulent boundary-layer flow on portions of wings, nacelles, tails and fuselages. Methods to apply robust micro-scale 'riblet' geometries for turbulent-flow skin-friction reduction continue to be developed and tested to progress maturation to practicality. Estimates suggest opportunities of 1–2% fuel-burn reduction on new and existing aircraft with significant areas covered by practical 'riblets'. A more significant reduction in skin-friction drag is possible by maintaining laminar flow on forward areas of engine nacelles, wings and tails. Surfaces intended for Natural Laminar Flow (NLF) are already present on some in-production commercial and business-jet aircraft (e.g. nacelle-inlet lip and winglets on some larger aircraft and portions of wing and fuselage on some business jets). Achieving laminar flow on aircraft requires well-balanced aerodynamics, structural designs and aligned manufacturing methods to meet necessary surface quality. Overall, the practical and robust achievement of significant laminar flow on wings and other surfaces could reduce aircraft fuel burn by up to 5%. The magnitude of potential benefit depends on the fraction of airplane surfaces manufactured to achieve laminar flow. Inevitably, advanced manufacturing technologies to meet new design requirements are complex. As such, the need for advanced materials, special tooling and technics might add to the environmental footprint and economics at the production stage.

Because air is a viscous medium, a body subjected to a moving airstream will inevitably have, through viscous adhesion, a thin layer of air at its surface with zero relative velocity. Through the same action, the succeeding layers with increasing distance from the surface will be subject to retardation but to a lower degree. A point is, therefore, reached where the airflow will be unaffected and its velocity will be that of the 'free stream' airflow. In conventional airfoils, the laminar boundary layer extends up to 0.07 inches from the surface wall, while the turbulent part is up to 0.7 inches. The boundary layer is a thin air layer from the surface where the air molecules, at zero velocity, reach the point where there is no retardation. Usually, the boundary layer velocity flow is less than 99% of the free stream value.

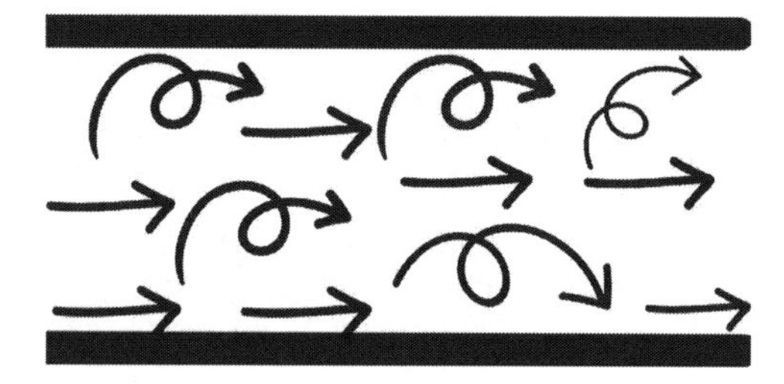

*Figure 5.2* Laminar vs turbulent flow

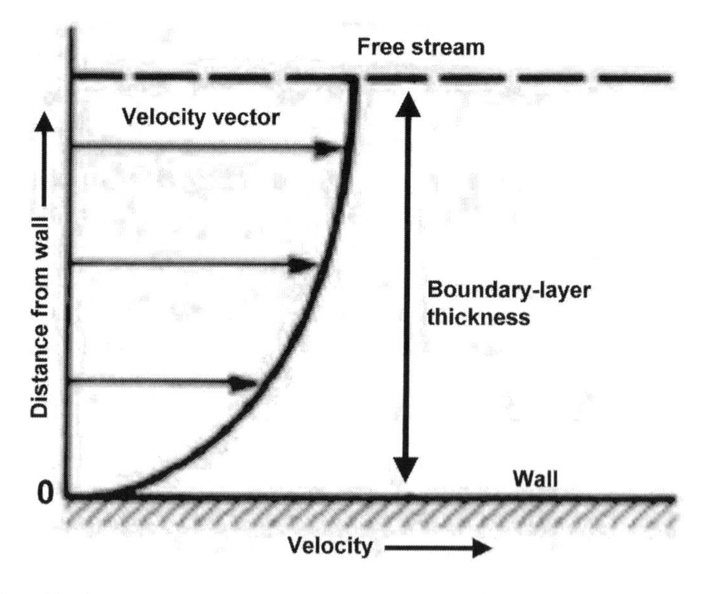


*Figure 5.3* Boundary layer

Natural laminar flow is dependent on the body shape or the airfoil shape. The inherent shape design helps the laminar flow to stay attached further over the body or airfoil than the conventional shapes. The NLF benefits are restricted to moderate swept wings, but research efforts are trying to extend this principle to high swept wings cable reaching the transonic range (between Mach 0.8 and 1.2). Although NLF research focuses on wing areas, recent research includes other aircraft areas such as fuselage, nacelles, engine lip-skins, etc.

Laminar Airflow and Laminar Flow Control (LFC) technology tries to keep the laminar boundary layer attached to a large part of the wing by virtually 'sucking' the turbulent boundary layer or reenergising the laminar flow before separation. Hybrid LFCA or HLFC is a variation of LFC that uses NLF for a more extensive portion of the wing to reduce the power required to eliminate the turbulent boundary layer growth. The main weakness of such techniques is that they are active rather than passive. In other words, additional systems should be in place to transfer energy.[1] Due to relatively low price of aviation fuel at the 60s and 70s and the practical application difficulties, Laminar Flow Control (LFC) research was discontinued at the end of the 1960s. Nowadays, this research has been revived in response to the increased fuel prices and the growing awareness about the need of lower engine emissions reflecting to the environment. The LFC technology has the potential for considerable improvement in fuel economy. For transport aircraft, the reduction in fuel burnt could reach up to 30%. Although the active LFC systems add weight and cost and their complexity requires more maintenance, recent manufacturing technics have assisted to overcome some of the practical application difficulties.[2]

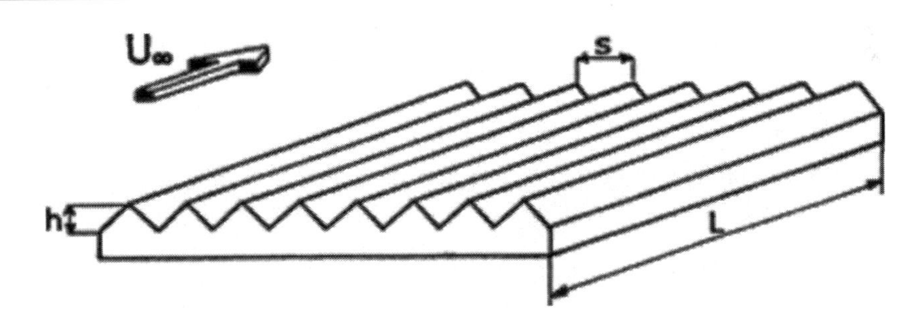


*Figure 5.4* Micro-scale riblet

While in traditional aerodynamics Laminar Airflow is always desirable, practically, it is impossible to achieve. Different flight conditions and surface imperfections result in unwanted transitions from laminar flow to turbulent. There were several empirical attempts to control turbulent flow or even use it for the benefit of laminar (i.e. vortex generators of late 50s high-speed aircraft). Even when the laminar flow is maintained over a significant percentage of the wing, most of the aircraft surface (i.e. fuselage, engine cowlings) are subject to turbulent flow. However, the observation of nature gave the engineers solutions to minimise skin friction drag in a turbulent flow regime. For example, the skin of fast-swimming sharks reveals riblet structures aligned in the direction of flow. Experimental evidence in wing tunnels confirmed that riblet surfaces reduce skin friction drag in a turbulent flow. Riblet surfaces are tiny streamwise grooved surfaces (grooves aligned with the flow). Many riblet shapes have been tested; some are tested now in several research projects. As an example, for triangular shape with $h=s$, a friction drag reduction of 8% is obtained when $h^V = S^v$ *is* in the range of $v=10–15$. These figures give a few hundredths of a millimetre in terms of physical dimensions.[2] Covering certain surfaces of an aircraft with riblet film or even embedding such surfaces in skin manufacturing shows promising results in total drag reduction. Still, manufacturing challenges due to microscopic dimensions and the associated costs limit their use to newer designs.

***Propulsion***

Three technology paths can reduce propulsion-system fuel consumption: increase thermal efficiency by increasing the compressor Overall Pressure Ratio (with consequent increase in core engine operating temperatures); increase propulsive efficiency by increasing the engine Bypass Ratio (*BPR*) and, consequently, fan diameter; and reduce installed engine weight and drag. Over the last decade, newly introduced aircraft and major derivatives with new engines have followed these paths as diameters of engines have increased, while aircraft manufacturers have maintained acceptable installation and integration penalties. Between 2016 and 2023, advanced technology engines have entered or will enter service on new and reengined aircraft. New technology engines at *BPR*'s 9 to 12 for regional jets and single-aisle aircraft (such as the E2, A220, A320neo, B737MAX, MRJ, MC-21 and C919) provide a significant 15% reduction in fuel burn relative to earlier *BPR*~5 engines. Latest generation engines for new production twin-aisle aircraft (A330neo and B777–9) can deliver 10% fuel-burn reduction relative to 2014 in-service reference.[3]

***Structure and Materials***

A key opportunity to reduce fuel burn and CO2 emissions is further minimisation of aircraft structural weight. Reduction in empty weight while maintaining structural requirements (strength, stiffness and safety) may be done with several levers:

- further optimisation of established structural technologies and/or materials;
- introduction of new materials and/or structural technologies; and
- alternate aircraft architectures.

Composite materials and structures technology have been developed and introduced in several new small and large aircraft. There is still progress anticipated in allowable margins linked to existing materials and in new designs targeting improved assembly process (such as bonding, stitching and welding). Aircraft manufacturers recognize the individual advantages of composites and advanced metallic alloys – and aim for optimum balance of both materials. For metallic materials, new alloys have been developed to be competitive with composites for thin parts applications (such as fuselage skins). Overall, future weight-reduction opportunities derived from combination of described technologies is estimated to be as much as 8% relative to current state-of-the-art structural configurations.[3]

**What Is Life Cycle Assessment?**

The average life cycle of an aircraft from its purchase to retirement is about twenty to thirty-six years depending on the model. The concern is big about the retiring aircraft and the environmental implications of decommissioning aircraft in all around the world. The industry is in a process of adopting sustainable design of aircraft, with a sustainable use through fuels, best practices and operational efficiency until its decommissioning. This is where the life cycle process and assessment come in the way.

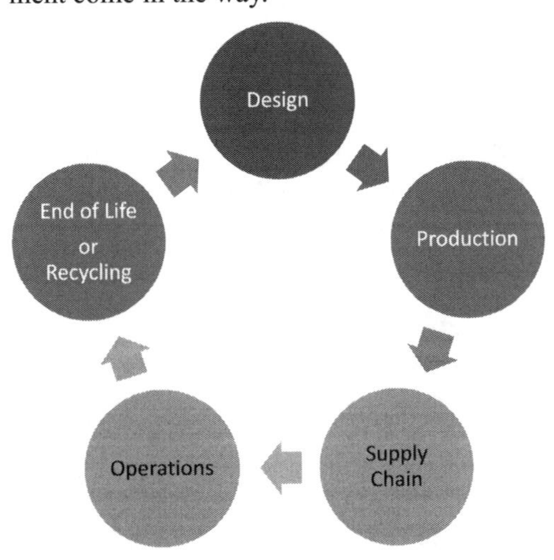


*Figure 5.5* The five basic phases of aircraft cycle

Figure 5.5 demonstrates the five basic phases of an aircraft life cycle. Carbon footprint and assessment should be done throughout all the five phases and their processes. Nonetheless, in Life Cycle Assessment, aviation should show a great responsibility for the end products and services with an increased stewardship. To achieve that, the aviation industry should consider that product stewardship involves a strategy for ensuring the proper sourcing, design, production, distribution, use and end-of-life stages of an aircraft or a part. This concept relies on efforts to minimise their footprint by performing life cycle assessments on products and services they offer. As a general description, life cycle assessment projects are designed to assess life cycle environmental impacts of products or projects using standardised procedures and metrics. LCA is a powerful tool that combines systems thinking with a quantitative approach. It is based on the holistic tracing of inputs (material and energy) and outputs and the normalisation of those outputs to various 'impact categories'. The scope of an LCA study often varies from 'Cradle to Grave' to focus on different segments of a product life cycle 'Cradle to Gate', 'Gate to Gate', or 'Gate to Grave', where 'Gate' can represent almost any stage of the life cycle assessment.

> Life Cycle Assessment is a systematic and standardised approach. Its purpose is to quantify the potential environmental impacts of a process or a product, starting from raw materials extraction to its end of life.

An LCA is not always required to substantiate product claims; however, it is required for Environmental Product Declarations (EPD), which is a highly structured approach for communicating life cycle data. An EPD is a document that aims is to communicate potential environmental impacts of a product, from cradle to grave.

In 2016, Bombardier CS 100 aircraft received the first environmental product declaration. The EPD, which is based on verified Life Cycle Analysis (LCA) data, discloses information about the life cycle environmental impact of products and thus provides the basis for a fair comparison of products and services regarding their environmental performance. For this life cycle process, 80% of the total environmental impact is defined at the design stage. The final aircraft assembly line and wing manufacturing sites were certified under LEED. The materials used made the aircraft structure lighter, and the brakes are electric, differentiating from the traditional hydraulic. It is claimed by the manufacturers that this aircraft has the lowest fuel burn per passenger and the lowest emissions and noise levels. This a practical example shows how LCA can work in all stages of aircraft – from design to its use, reducing its environmental footprint from the early design stages.

In a general scope, LCA relies on databases and institutions that store information about different types of processes and products. Thus, life cycle studies are rarely able to rely purely on collected or 'primary' data. These secondary data sources typically rely on data that has been averaged across multiple locations to present 'regional industry averages'. However, large differences can result from local variations, such as the carbon intensity of an electrical grid. Therefore, it is best to obtain localised data to the maximum extent feasible. LCA works extremely well for identifying 'hotspots' for more detailed analysis in a specific value chain. It also helps substantiate communications related to specific impacts while in the product development stage when materials and production process are being identified. According to the Sustainability Consortium, an NGO dedicated to improving the sustainability of consumer products, over 2,500 surveys involving over 1,700 suppliers indicate that 'most manufacturers have limited visibility into their supply chains and their related sustainability risks'.[20] This illustrates the need for specific life cycle approaches and procedures. LCA methodologies are standardised through the International Standards Organization's ISO 14040 series of standards that 'give guidelines on the principles and conduct of LCA studies'.[6]

Life Cycle Assessment examines the environmental impact of all stages of a product life. A typical LCA, at a minimum, addresses environmental impacts:

- global warming potential (GWP)
- abiotic depletion potential (ADP)
- acidification potential (AP)
- eutrophication potential (EP)
- ozone depletion potential or ozone depleting chemicals (ODC)
- photochemical ozone creation potential (POCP)

Of course, when we talk about aviation, we might want to go deeper and be more specific on the environmental effects – an analysis we did in Chapter 3. Each one of these impact categories is examined separately, as it relies on the final product, the type of resources, the technical characteristics of each indented use, the location where the product is developed, etc. In addition, for some cases, there is not enough information for each life cycle phase and that makes the whole process a lot more difficult to analyse per product and operation. Since LCA is a valuable part of the acceptable sustainability practices, economic and social conditions must be considered. The UNEP'S report 'Social Lifecycle Assessment Guidelines' includes assessment of impacts on the five stakeholder categories – workers or employees, local community, society at large, consumers and value chain actors. One life cycle assessment system combines environmental and social impacts in a more qualitative rather than strictly quantitative manner, allowing categories such as national security, worker health, local economic impacts, etc., to be assessed and options compared using a points system. The U.S. EPA suggests inclusion of environmental justice issues in environmental impact studies. All this various information are essential since they support the creation of a valid and industry-accepted life cycle inventory – a necessary LCA element for all products, their use and the life cycle phases. As in all industries, aviation has its own databases and organisations that can provide similar information; however, it's more likely to see different manufacturers to create their own life cycle system and inventories and then share the information. ICAO offers a great set of information about LCA; however, it focuses on the life cycle of the creation of sustainable fuels.

The four phases of a life cycle assessment under UNEP:

- **Goal and Scope Definition** – Define the product(s) or service(s) for assessment; choose a functional basis for comparison.
- **Inventory of Resources** – Quantify for each process: the inventory analysis of extractions and emissions, the energy and raw materials used and the emissions to the atmosphere, water, and land. Then they are combined in a process flow chart created for the scope, and they are compared to the functional basis.
- **Impact Assessment** – Assess the effects of the resources used and their emissions. Then the effects are measured and arranged into a smaller number of impact categories.
- **Interpretation** – The results are reported in an informative way, identifying the need and opportunities for reducing the environmental impact of the defined product(s) or service(s).

## ISO Series Standards 14040 and LCA[8]

### *ISO 14040*

The definition of LCA uses two definitive words '*systematic*' and '*standardised*' approach. These two words are signifying that this approach should follow a very specific methodology and format, through the standardisation offered by the ISOs. Hence, there is a cluster of ISOs that are part of the environmental ISO Group of 14000 and, specifically, the ISO 14040 series standards, applicable to the aviation and aerospace. Life Cycle Assessment (LCA) is best known for the quantitative analysis of a product's environmental aspects over its entire life cycle. Products in this context include both products and services. Emissions to the air, water, and land such as CO2, biochemical oxygen demand (BOD), solid wastes and resource consumptions constitute environmental loads. Environmental impacts in the LCA context refer to adverse impacts on the areas of concern such as ecosystem, human health and natural resources. There are four phases in an LCA: goal and scope definition, life cycle inventory analysis, life cycle impact assessment and life cycle interpretation. The ISO 14040 series standards, Life Cycle Assessment, address quantitative assessment methods for the assessment of the environmental aspects. ISO 14040 is an overarching standard encompassing all four phases of LCA. There are three more standards supplementing ISO 14040. ISO 14041 deals with goal and scope definition and life cycle inventory methods. ISO 14042 deals with life cycle impact assessment methods and ISO 14043 life cycle interpretation methods. ISO 14044 specifies requirements and provides guidelines for Life Cycle Assessment (LCA).

a) the goal and scope definition of the LCA;
b) the life cycle inventory analysis (LCI) phase;
c) the life cycle impact assessment (LCIA) phase;
d) the life cycle interpretation phase;
e) reporting and critical review of the LCA;
f) limitations of the LCA;
g) relationship between the LCA phases; and
h) conditions for use of value choices and optional elements.

Setting the ground of ISO 14040 and their implementation to aviation, it's necessary to develop first the baseline of LCA. The questions that need to be asked: why do we need an LCA, what is the goal of LCA and what kind of benefits should be expected?

***Life Cycle Assessment Methods***

As mentioned earlier, LCA is a method to study the environmental footprint of a product at different phases of its life cycle. There are several approaches of LCA: cradle-to-grave approach, cradle-to-cradle approach, cradle-to-gate, gate-to-gate, well-to-wheel or gate-to-grave. Depending on the product, these methods can be selected and define alternative use of the LCA results. A cradle-to-grave LCA approach requires a significant set of data that represent the life cycle inventory (LCI). The inventory collects all the information that are necessary to calculate the LCA. The LCI is possibly the most important and difficult step of LCA journey. In the case a complete aircraft LCA, the LCI is created by the amount and type of raw materials necessary to its structure (i.e. carbon and glass fiber, aluminium, nickel, titan, etc.), types of energy consumed (i.e. electricity, heat energy) and also chemical compounds (i.e. CO2, NOx, CO, etc.) that may be emitted to the air, ground or water throughout the projected life. Another issue is the change in land use (i.e. airports, assembly lines, refineries, etc.) The goal of LCA is to identify the phases that have the greater environmental footprint. Those phases are evaluated and analysed during the design stage, aiming to reduce fuel consumption (and emission) from the aircraft use. At the same time, the objective is to reduce the environmental footprint as much as possible during manufacturing. Under the same perspective, Sustainable Aviation Fuels are developed. The LCA is a method that is useful since it can indicate processes with increased carbon production, giving the opportunity to re-evaluate the processes and their environmental footprint.

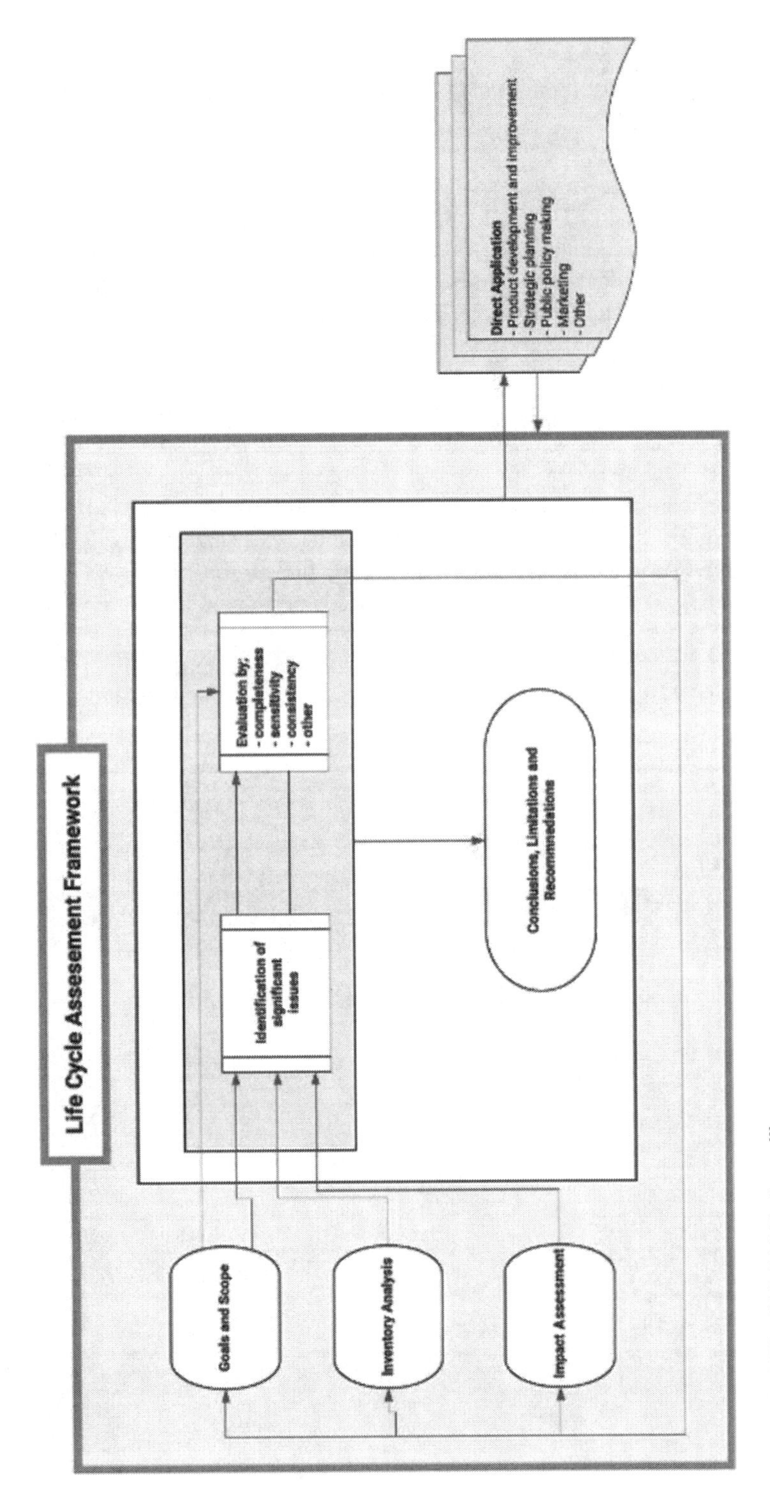


*Figure 5.6* The ISO 14040 phases[9]

- **Cradle to Grave:** It is a life cycle method that measures environmental impacts at all stages of a product's life cycle, from how natural resources are extracted from the ground and processed, to each subsequent stage of manufacturing, transportation, product use and, ultimately, disposal.
- **Cradle to Cradle:** It is the life cycle method where a product is developed and used. At the end of its life, it is fully recycled, imitating nature's cycle, leaving no waste.
- **Cradle to Gate:** This method refers to all the initial steps of the development of the product until the moment it is shipped to the client who ordered that product and it is ready to use. A production/manufacturing company can measure cradle-to-gate because they've designed a product that can be easily recycled, avoiding the landfill.
- **Gate to Gate:** This method is part of the cradle-to-gate method, and it refers to processes between product and use phase.
- **Gate to Grave:** This process is part of the cradle-to-gate as well, and it refers to the processes from the production phase to end use or recycling of the product.
- **Well to Wheel:** This approach applies only to vehicles, from cars to aircraft. It refers to fuels and the materials and resources extracted to produce the fuel until the delivery and consumption of it. The well-to-wheel approach has two stages: i) the well-to-station stage that involves the extraction or production of fuel to the 'upstream' section, and ii) the station-to-the-wheel that involves how fuel is consumed and burned, sometimes also referred to as the 'downstream' part. The well-to-wheel analysis assesses the energy consumption of various fuel transportation means until fuel is at the pumping stations. As described earlier, the process of creating Sustainable Aviation Fuels relies greatly to that LCA method.

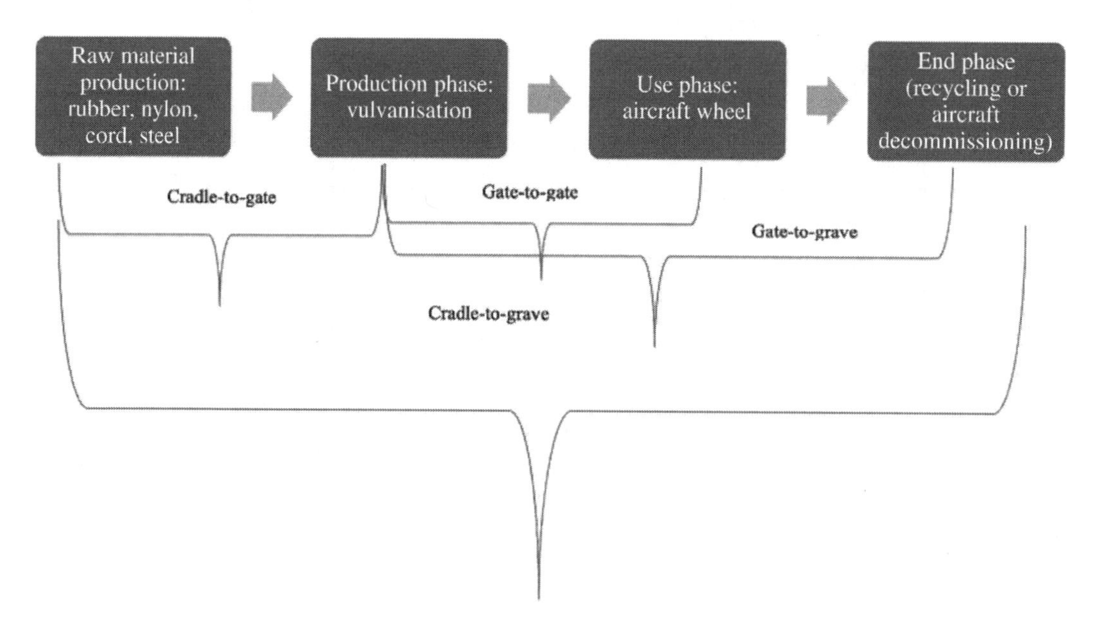


*Figure 5.7* Example of LCA method in an aircraft tire

Closing this section, it is essential to highlight that life cycle assessments are rarely able to rely purely on collected or 'primary' data. Currently, this is one of the main weaknesses of LCA mainly because there are not always reliable databases available for all existing processes and products developed. Therefore, secondary data sources are frequently used. These secondary data sources typically rely on data that has been averaged across multiple locations and industries and industrial disciplines. The basic role and benefit of LCA is that it identifies 'hotspots' that need more detailed analysis in a specific value chain process. It can define where environmental footprint needs to be reduced either by altering the process, the energy consumed or even the type of primary resources. It also helps specify impacts during the product development stage, when materials and production processes are identified.

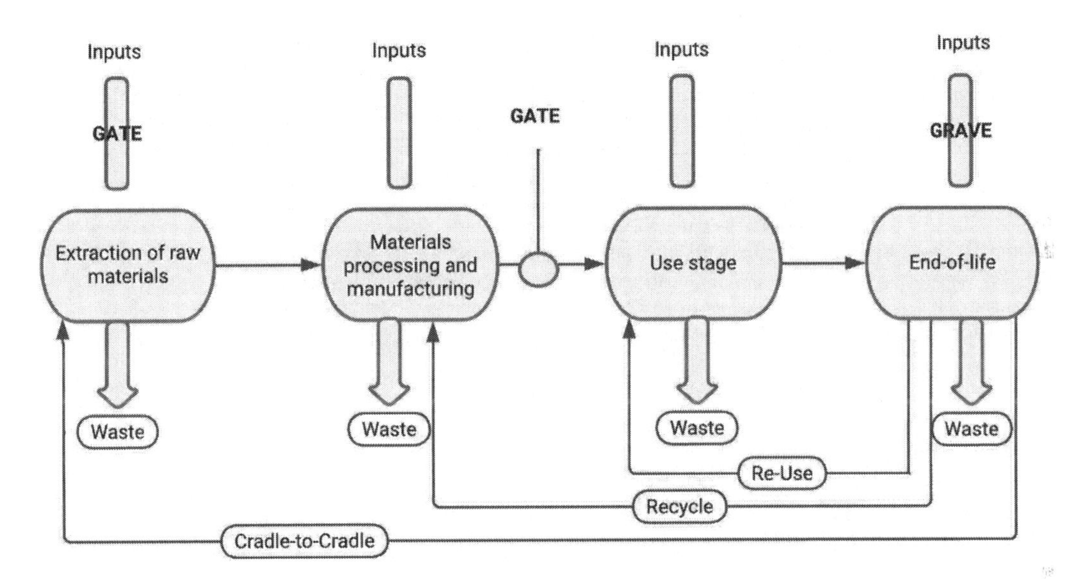


*Figure 5.8* Ecological loop[9]

## Conclusion

In this chapter, we have examined the role of the design of aviation products in achieving the sustainability goals. Aviation industry had always been a sector that pursued technological solutions to complex problems. The introduction of sustainability principles to the new aircraft designs made visible what the industry had been trying from its early steps. Flying higher, faster and in longer distances had been always the ultimate goal of aviation. However, nowadays, this notion embraces in a systematic manner the concerns of the environmental impact of such designs. As the aviation sector is expected to grow, new propulsion systems, more efficient engines and improved aerodynamic performance aircraft offer significant reductions in CO2 emissions. The entry into service aircraft like Boeing 787-9 and -10, the Airbus A350-900 and -1000 and the refitting of new composite wings to Boeing B777-9 are some examples of design enhancements offering savings up to 2% of fuel burn. The effort, however, never stops. New technologies like micro-scale riblet are tested to control the turbulent flow which if combined with natural flow control show promising results. Old technologies like the Laminar Flow Control are revisited using advance manufacturing technics and improved materials. High Bypass Ratio engines with increased thermal efficiency and Overall Pressure Ratio installed to new designs and older variants are refitted. The new propulsion systems deliver 10% fuel-burn reduction. However, one of the most critical tools during the design phase is the Life Cycle Assessment.

The LCA is designed to assess life cycle environmental impacts of products or services using standardised procedures and metrics. LCA is a powerful tool that combines systems thinking with a quantitative approach. It is based on the traceability of material and energy used as inputs and the normalisation of outputs to various 'impact categories'. The role of ISO 14000 series standards 14000 and, in particular, the ISO 14040 analysed and how it contributes to the '*systematic*' and '*standardised*' applicability of the LCA methodology. The scope of an LCA study was explained presenting the basic methods: 'Cradle to Grave', 'Cradle to Gate', 'Gate to Gate' or 'Gate to Grave' when the focus is on different segments of a product life cycle.

*Table 5.1* Acronym rundown

| | |
|---|---|
| LCA | Life Cycle Assessment |
| NLF | Natural Laminar Flow |
| LFC | Laminar Flow Control |
| HLFC | Hybrid Laminar Flow Control |
| BPR | Bypass Ratio |
| EPD | Environmental Product Declaration |
| GWP | Global Warming Potential |
| POCP | Photochemical ozone creation potential |
| ODC | Ozone depletion potential or ozone depleting chemicals (ODC) |
| EP | Eutrophication potential |
| ADP | Abiotic depletion potential |
| AP | Acidification potential |
| BOD | Biochemical Oxygen Demand |

**Key Points to Remember**

- Improvements in aerodynamic, propulsion and structures technologies have a direct link to aircraft emissions reduction. Improvements in systems design and manufacturing technology are also key to achieving future aircraft CO2-reduction goals.
- New aerodynamic designs may reach 1–2% fuel-burn reduction on new and existing aircraft with significant areas covered by practical 'riblets'.
- More reduction in skin-friction drag is possible by maintaining *laminar* flow on forward areas of engine nacelles, wings and tail
- The main drawback of an active Laminar Flow Control systems is the additional weight and complexity.
- Natural laminar flow is dependent on the body shape or the airfoil shape.
- Combination of composites and advanced metal alloys in airframe structures present future weight-reduction opportunities estimated to be as much as 8%
- Micro-scale riblet surface has tiny grooves aligned with the flow.
- The average life cycle of an aircraft from its purchase to retirement is about twenty to thirty-six years depending on the model.
- Life Cycle Assessment is a systematic and standardised approach. Its purpose is to quantify the potential environmental impacts of a process or a product, starting from raw materials extraction to its end of life.
- The aviation industry should consider that product stewardship involves a strategy for ensuring the sourcing, design, production, distribution, use and end-of-life stages of an aircraft. This concept relies on organisations and their effort to minimise their footprint by performing life cycle assessments on products and services they offer.

- Life cycle assessment projects are designed to assess life cycle environmental impacts of products or projects using standardised procedures and metrics. LCA is a powerful tool that combines systems thinking with a quantitative approach. It is based on the holistic tracing of inputs (material and energy) and outputs and the normalisation of those outputs to various 'impact categories'.
- The scope of an LCA study often varies from 'Cradle to Grave' to focus on different segments of a product life cycle 'Cradle to Gate', 'Gate to Gate' or 'Gate to Grave', where 'Gate' can represent almost any stage of the life cycle.

**Chapter Review Questions**

5.1 Describe the laminar boundary layer.
5.2 What are the basic Life Cycle Assessment stages?
5.3 How do you explain the term 'sustainable design' for an aviation product?
5.4 Describe the stages of LCA for an aircraft tire.
5.5 Considering the fifteen out of 17 SDGs for aviation, provide an explanation on which of them can be covered from the aviation industry through the implementation of Life Cycle Assessment. To answer this question, you might need to select a specific aviation product where LCA is applied.

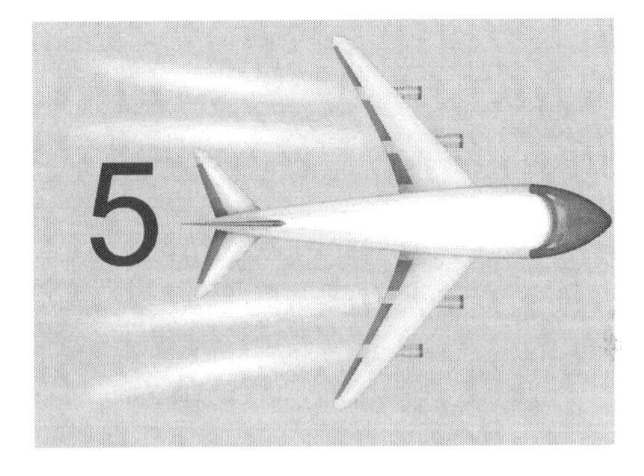


*Figure 5.9* Airplane no. 5 pointing right

5.6 How does LCA contribute to the other two pillars of sustainability, except of the environmental pillar?

**References**


[1] Gudmundsson, S. (2014). *General aviation aircraft design: Applied methods and procedures* (1st ed.). Butterworth-Heinemann

[2] Houghton, E. L., Edward, L., Carpenter, P. W., Collicott, S. H., Valentine, D. T., & Carpenter, P. W. (2017). *Aerodynamics for engineering students* (7th ed.). Butterworth-Heinemann

[3] International Coordinating Council of Aerospace Industries Associations. (2019). *Advancing technology opportunities to further reduce CO2 emissions*. ICAO. www.icao.int/environmental-protection/Documents/EnvironmentalReports/2019/ENVReport2019_pg116-121.pdf

[4] ATAG. (2016, September 27). *Bombardier CS100 aircraft receives the first environmental product declaration in the industry*. Aviation Beyond Borders. https://aviationbenefits.org/newswire/2016/09/bombardier-cs100-aircraft-receives-the-first-environmental-product-declaration-in-the-industry/

[5] Claro, M. G. (2017). *The CS100 EPD® the first-ever environmental product declaration in the aerospace industry*. Bombardier. www.utias.utoronto.ca/wp-content/uploads/2017/09/Garcia-Claro-CS100EPD.pdf

[6] European Environment Agency. (n.d.). *cradle to grave*. www.eea.europa.eu/help/glossary/eea-glossary/cradle-to-grave

[7] Klemeš, J. J. (2015). *Assessing and measuring environmental impact and sustainability* (1st ed.). Butterworth-Heinemann

[8] ICAO Secretariat. (2019). *Environmental report 2019 aviation and environment*. ICAO. www.icao.int/environmental-protection/Documents/EnvironmentalReports/2019/ENVReport2019pg228-231.pdf

[9] ISO. (2014, August 12). *ISO 14044:2006*. www.iso.org/standard/38498.html

[10] Chandel, A., & Silveira, M. H. L. (2018). *Advances in sugarcane biorefinery: Technologies, commercialization, policy issues and paradigm shift for bioethanol and by-products*. Elsevier

# 6 Sustainable Aviation Fuels

### Chapter Outcomes

At the end of this chapter, you will be able to do the following:

- Identify the different aviation fuel types.
- Understand what Sustainable Aviation Fuels (SAF) are.
- Explain the different processes for the SAF production.
- Explain the role of ICAO CORSIA and the use of SAF.
- Align the use of SAF with the Sustainable Development Goals.

## Introduction

In October 2021, the aviation world decided to participate in the effort to meet the net zero goals by 2050. Sustainable Aviation Fuels (SAF) is currently the most popular way to meet these goals. However, the capacity of SAF production is unable to cover the airlines' demand internationally. The SAF quantities must be substantial and readily available to refueling locations globally so that this effort can have a positive outcome in decarbonising the sector. We often hear in the news that airlines made 100% SAF flights, yet these flights are only test ones. Safety and economic factors are always key drivers behind every major shift in the aviation sector. As such, it appears that currently, the industry has a long road until the use of SAF reaches 100% in the commercial flights, but this effort seems very promising. Up today, there are eleven processes for producing certified types of SAF. For each different certified SAF type, there is a different blending ratio. ASTM is the organisation that certifies different types of SAF with a blending ratio of SAF and conventional fuel. The most common conventional fuel used are the following: Jet A1, or kerosene, is light-refined petroleum; Jet B is of similar chemical structure but for low temperatures; and AvGas 100LL is a low-lead kerosene. The 'sustainability' of Sustainable Aviation Fuels is based on the way SAF is produced, the feedstocks used as a primary source and the way they are extracted and processed. A SAF's footprint during its production – from the time the primary source is extracted until the fuel is developed – should be traceable throughout the entire production life cycle. At the same time, the SAF should be able to meet all the sustainability criteria apart of reducing the flight emissions. The industry is leaping a different pathway, yet the environment should be one of many focuses. Economic and social considerations can and must also be part of this process while developing new fuel technologies.

## Aircraft Fuels

Aviation fuels are used for aircraft propulsion by converting their chemical energy to thermal energy via combustion and, eventually, to kinetic energy via extraction of burned gases. The phrase 'aviation fuels' has caused much controversy and arguments about aviation's position in

DOI: 10.4324/9781003251231-6

environmental sustainability in the past decade. The necessity to seek less harmful aviation fuels leads to developing new alternatives, more environmentally friendly with less emissions, and more sustainable practices of primary resources extraction, production and distribution. This is what we call Sustainable Aviation Fuels (SAF). Before discussing this new entry, let us briefly overview the conventional aviation fuels used for decades. Knowing where we are and our status is essential to find the most appropriate and efficient solution for the future.

- **Jet Fuel – Jet A1:** The Jet A1 type of aviation fuel is the most common fuel used globally in jet engines and turboprops in the commercial aviation. The substance of Jet A1 is light-refined petroleum, and the fuel type is kerosene. Jet A1 has a flash point – the temperature at which a particular organic compound gives off sufficient vapour to ignite in the air – higher than 38°C, and its freezing point is −47°C. Jet A is a similar kerosene type of aviation fuel. It has a higher freezing point and is mainly used in the United States.
- **Jet B – kerosene-gasoline mixture:** Jet B is used in military jet engines and is a blend of 35% kerosene and 65% gasoline. This type of fuel has a flash point of 20°C and a freezing point even below −72°C, and it is used in regions with particularly low temperatures. It should be noted that the aircraft engines must be certified for Jet B use.
- **Aviation Gasoline AvGas:** This fuel is used mainly in older piston engines of sport airplanes and single-engine private aircraft, which require leaded fuel with high octane numbers. AvGas 100LL is the only avgas that meets these requirements since, essentially, it is low-lead gasoline.

The newest aviation fuel type is the sustainable alternative or aviation fuel. It should be noted that the SAF is used in a mixture of conventional aviation fuel types. The use of SAF as aviation fuel has made a significant entrance, and eventually, it will gain more ground when feedstock and other issues are solved.

The SAF is not like Jet A fuel, it is Jet A!

**Facts:**

- Aviation jet fuel accounts for 2.5% of total global CO2 emissions
- The combustion of 1 kg of jet fuel in an aircraft engine produces 3.16 kg of CO2.
- The global aviation industry produces almost 2.1% of all human-induced CO2 emissions.
- Aviation is responsible for 12% of CO2 emissions from all transportation means, compared to 74% of road transportation.

### Sustainable Aviation Fuels

Sustainable sources and natural feedstocks are the primary sources of producing Sustainable Aviation Fuels (SAF). It is notable to say that they have very similar chemistry to traditional jet fuel. Using SAF, results in fewer carbon emissions compared to what jet fuels emit and less emissions over the fuel's life cycle. Sustainable Aviation Fuels (SAF) is the term used by the aviation industry, given that this term is broader to aviation biofuels. The term 'biofuels' generally is used to explain fuels generated from biological resources, like plants or animal material. However, current aviation technology uses fuels from other alternative sources, often non-biological. Hence, the term SAF is adjusted to highlight the sustainable nature of these fuels. Noteworthy is that presently, SAF is blended with conventional kerosene. Because of this blending, the final product for use has a fossil-based chemical structure blended with renewable hydrocarbon. Then it is certified as 'Jet A1' fuel and can be used without requiring technical modifications on engine or aircraft fuel systems to be an aircraft fuel.[1]

Sustainable Aviation Fuels consist of three key elements as shown in the acronym:

- **Sustainable:** The concept of sustainability requires a set of actions and practices that will support the three pillars: environment, economy and society. Sustainability in aviation fuels means producing less environmental effects during their consumption and production, with fewer emissions in the entire life cycle processes. The primary location, the way the feedstock is cultivated and the involved personnel and how they are treated during the cultivation are also considerations of the SAF sustainability.
- **Alternative:** Alternative fuels are considered any other than conventional which are produced by carbon-based sources such as oil, gas, coal, etc. The word alternative means that this fuel type is differently processed from jet fuel. The primary sources of SAF vary from cooking and plant oils to municipal waste and agricultural residues. More information about feedstock to create SAF is given in the following sections of this chapter.
- **Fuel:** Any source that is used to produce energy is called fuel. In aviation, the term 'fuel' signifies the appropriate type of jet fuel with all the technical and certification requirements for safe use and operations in commercial aircraft. As of the International Civil Aviation Organization (ICAO), the term 'alternative fuels' is defined as 'any fuel that has the potential to generate lower carbon emissions than conventional kerosene on a life cycle basis'.[2]

The emissions during the production of SAF come from equipment used to cultivate the crop, transport the feedstocks and refine the fuel, to name a few. Sustainable Aviation Fuels can provide a significant reduction in overall CO2 emissions, which can be up to 80% compared to fossil fuels. Furthermore, using SAF can significantly reduce sulphur, sulphur dioxide and other particulate matter emissions. Feedstocks to create SAF are processed through chemical processes with electric energy, water and CO2. However, the feedstocks can be both sustainable and unsustainable, depending on the methods used to extract them. That is why the LCA of the feedstock production and the methodologies to make the fuels distinguish biofuels or biobased fuels from sustainable fuels. The difference between the two types of fuels – biobased and sustainable fuels – lies in how they are produced, what the primary resources are and how they are extracted and processed. This is why the aviation industry must follow strict, independently-verified sustainability standards; hence, fuels must be appropriately certified to be used on commercial flights.

*Figure 6.1* Aircraft contrails

With respect to greenhouse gas emissions (GHG) under CORSIA, SAF should achieve life cycle emissions reduction of at least 10% compared to a fossil fuel baseline of 89 grams of $CO_2$ equivalent per megajoule (g $CO_2$e/MJ). According to RED II, to qualify biofuels as renewable energy sources, fuels have to achieve a 65% greater reduction in emissions against a fossil fuel baseline of 94 g $CO_2$e/MJ.

The American Society for Testing and Materials (ASTM) International has developed standards to approve new biobased aviation fuels. There are eight production processes that have been certified for blending with conventional aviation fuel, and still, more are pending for approval. These include the following:

- **FT-SPK (Fischer-Tropsch Synthetic Paraffinic Kerosene):** Biomass is converted to synthetic gas and then into biobased aviation fuel. The maximum blending ratio is 50%.
- **FT-SPK/A:** It is a variation of FT-SPK, where alkylation of light aromatics creates a hydrocarbon blend that includes aromatic compounds. The maximum blending ratio is 50%.
- **HEFA (Hydroprocessed Fatty Acid Esters and Free Fatty Acid):** Lipid feedstocks, such as vegetable oils, used cooking oils, tallow, etc., are converted using hydrogen into green diesel, and this can be further separated to obtain biobased aviation fuel. The maximum blending ratio is 50%.
- **HFS-SIP (Hydroprocessing of Fermented Sugars – Synthetic Iso-Paraffinic kerosene):** Using modified yeasts, sugars are converted to hydrocarbons. The maximum blending ratio is 10%.
- **ATJ-SPK (Alcohol-to-Jet Synthetic Paraffinic Kerosene):** Dehydration, oligomerisation and hydroprocessing are used to convert alcohols, such as isobutanol, into hydrocarbon. The maximum blending ratio is 50%.
- **CHJ (Catalytic Hydrothermolysis Jet Fuel):** Triglycerides such as soybean oil, jatropha oil, camelina oil, carinata oil and tung oil are used as feedstock. The blending ratio is 50%.
- **HC-HEFA-SPK:** Synthesised paraffinic kerosene from hydrocarbon-hydroprocessed esters and fatty acids. Algae as feedstock. The blending ratio is 10%.
- **Co-processing:** Biocrude up to 5% by volume of lipidic feedstock in petroleum refinery processes.[2]

ASTM is a global leader in developing voluntary standards to be used by companies and individuals around the world. There are more than 12,800 standards. The ASTM manages jet fuel aviation specifications, and there are seven approved pathways with established criteria for blending Sustainable Aviation Fuels with conventional jet fuel. The role of ASTM is not to determine the sustainability of fuel but only to establish industry standards for its safety and performance.

***How to Produce Sustainable Fuels***[4]

As mentioned in the previous section, the there are numerous sources as feedstock to produce SAF. Current technology for SAF production allows sources such as municipal waste, used cooking oil and agricultural waste. In addition, crops and plants are also used for that same purpose.

- **Municipal solid waste (MSW):** Rather than dumping municipal waste in a landfill, adding to the development of CO2 concentration, it can be used to create sustainable jet fuel for aviation. The municipal waste comes from residential wastes, such as product packaging, old furniture, clothing, glass bottles, food scraps and used papers. The availability of municipal solid waste supply especially in large metropolitan cities shows great potential for using and processing it as a sustainable feedstock for SAF.

- **Cellulosic waste:** This type of waste comes from forestry, agricultural residues and excess wood. It can be processed through the FT-SPK process mentioned earlier. The biomass (wood or other agricultural waste) is converted into synthetic fuel or renewable isobutanol. Then it is further processed and converted to jet fuel.
- **Used cooking oil:** This source comes from the used animal or plant-based fat that has been used for cooking.
- **Camelina:** It is primarily an energy crop with high lipid oil content. Camelina oil is mainly a feedstock to produce renewable fuels. The remnants from the oil extraction can also be used as animal feed in small proportions. It is a fast-growing rotational crop with wheat and other cereal crops within the same year when the land would otherwise be left unplanted as part of the standard crop rotation program. Hence, it allows growers to diversify their crops and reduce mono-cropping. Mono-cropping has been shown to degrade soil, lower yields and resist pests and diseases. The supply chain of camelina can reach up to a 60% reduction in GHG emissions, meeting the requirement set by REDII.
- **Jatropha:** This plant produces seeds with a very high concentration of lipid oil used to produce fuels. Specifically, the seeds produce 30–40% of their mass in oil. A significant advantage of the jatropha crop is that it can be cultivated in non-arable areas, leaving space for edible or food crops. It is essential to mention that jatropha seeds are toxic to people and animals.
- **Halophytes:** This type of plant belongs to the saline habitat species, and they grow in areas of salt water or areas near sea spray.
- **Algae:** Algae is potentially the most promising feedstock for producing large quantities of SAF. These plants can be grown in polluted or salt water, deserts and other inhospitable places. They thrive off carbon dioxide, which makes them ideal for carbon sequestration. One of the most significant advantages of algae for oil production is the speed at which the feedstock can grow. It has been estimated that algae produce up to fifteen times more oil per square kilometre than other biofuel crops. Algae can be grown in areas that are not used for growing food, like the edges of deserts. However, algae cultivation has not fulfilled its early promise due to commercialisation challenges but continued research and development may result in the broader application of this feedstock in the future.
- **Non-biological alternative fuels**: This type of fuel production includes a 'power-to-liquid' process. It means creating jet fuel by involving electric energy, water and CO2. This fuel can be sustainable if the inputs are recovered as by-products of manufacturing and if renewable electric energy is used in production. For example, using the by-product waste gases of steel manufacturing to make Sustainable Aviation Fuels appears a great option. However, this process is costly and needs further research and development.

**Mono-cropping**: It is a type of cultivation where the production is generated only from one crop. This means that throughout the year, there will be periods where there will be no yield. While using a crop such as camelina as a source of sustainable fuel, the land can be used for the cultivation of another crop throughout the year, supporting the soil with nutrients that wouldn't exist when mono-cropping.

**Carbon sequestration**: It is a process that is used to capture and store carbon dioxide. There are three types of carbon sequestration: the biological, where CO2 is captured from natural systems, such as plants, grass trees, soils and oceans; the geological, where CO2 is captured and stored in geological formations; and the technological, which happens through the use of various innovative technologies.

The life cycle of SAF relies heavily not only in the processing of primary resources, meaning the crops, but also from the way the crops are cultivated. The actual name of this type of fuels signifies that sustainability must be the core element to be met. The difference between the conventional jet fuels and the Sustainable Aviation Fuels does not lay only in the final product but in the overall process of production until the final outcome of each process. That is why LCA is vital for the role of creating SAF, applying them in aviation and reducing the reliance in conventional fuels.

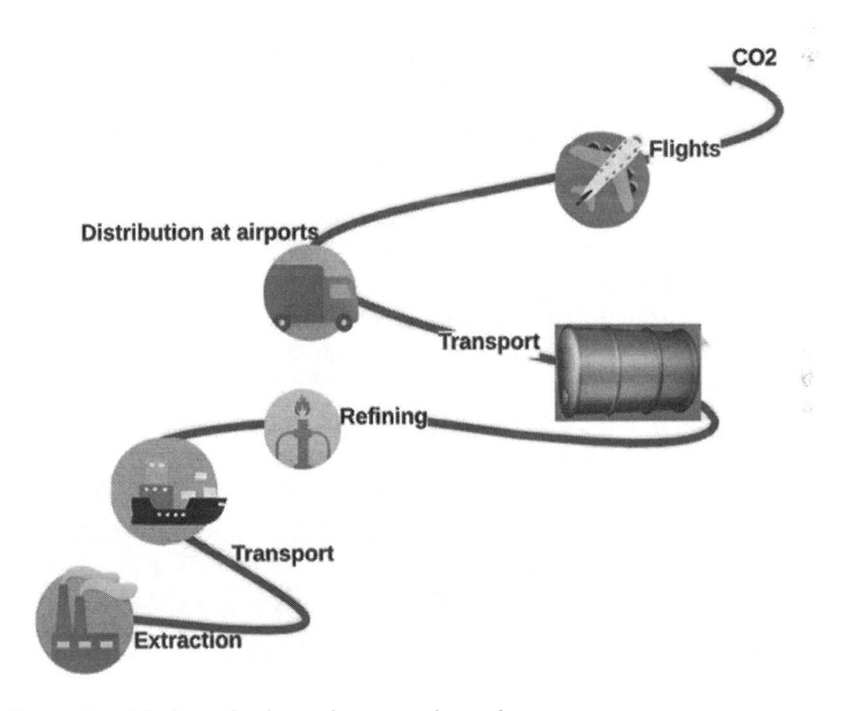


*Figure 6.2* Conventional fuels production and consumption pathway

Figure 6.2 follows a linear approach, a pathway with an end of processes. This means there is no circular approach in place. Therefore, reuse, recycle or systemic thinking approach must be implemented. Nowadays, the aim is always to 'start again in the cycle' pathway and return to where processes start so that a circular economy can bring both environmental and economic benefits. Hence, most of the fuel production and consumption pathway shown in Figure 6.2 needs an alternative. This alternative is shown in Figure 6.3

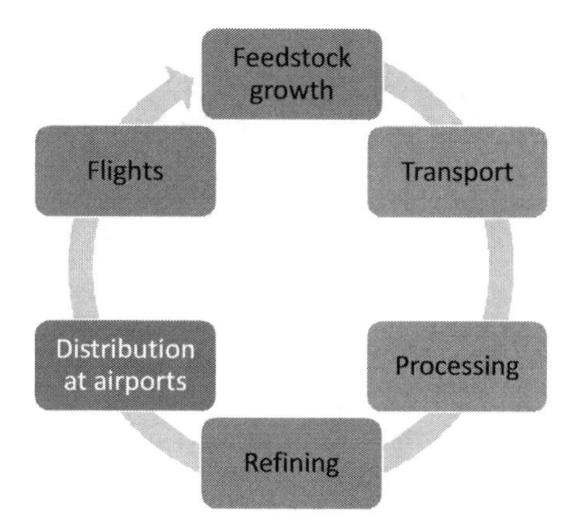


*Figure 6.3* Sustainable aviation fuels life cycle

## CORSIA and SAF

Chapter 3 discussed the functionality of the ICAO Carbon Offsetting Reduction Scheme for International Airlines (CORSIA) and its role in reducing aviation emissions. CORSIA supports the promotion and use of SAF in flights. However, to have fuels eligible to meet CORSIA requirements, they must meet certain criteria, with specific feedstock sources and a measurable life cycle emission value. Finally, an eligible fuel under CORSIA must be followed by a sustainability certification. The ICAO Assembly Resolution A39-3 requested the development of an appropriate methodology that air operators will use to calculate accurately how much their reduced carbon emissions are. This methodology will calculate, on an annual basis, their emissions while using sustainable alternative fuels under the provisions and basket measures reflected in the scheme. The starting point is the goal of the Paris Agreement to restrain the rise in global temperatures to less than 2°C, imposed as a necessity to the transportation sector. In addition, the sector is expected to cover more than 60% of the future oil demand. Since air travel has experienced growth and the projections are that it will continue to grow after the COVID-19 outbreak, ICAO has set a series of actions and measures to control the growth of CO2 emissions from the aviation industry. Among these actions is using Sustainable Aviation Fuels (SAF) as part of ICAO's basket measures to decrease GHG emissions.

The terms 'sustainable fuels' or 'alternative fuels' have been used interchangeably, but both demonstrate fuels made of non-conventional processes and resources with reduced environmental impacts. In CORSIA Annex 16, several definitions exist for these terms. The general definition is CORSIA Eligible Fuel. A CORSIA Eligible Fuel is a CORSIA Sustainable Aviation Fuels or lower-carbon aviation fuel. This fuel is appropriate for an operator to use to reduce carbon emissions. A CORSIA lower-carbon aviation fuel is an aviation fuel originating from fossil-based sources that meet the CORSIA Sustainability Criteria. Finally, a CORSIA Sustainable Aviation Fuel comprises renewable sources or waste and meets the CORSIA Sustainability Criteria.

***What Is a CORSIA Eligible Fuel (CEF) and How to Make It?***

So what is a CORSIA Eligible Fuel (CEF), and how is it produced? To identify that, we need to review the whole fuel supply chain – from the feedstock to the conversion process, to the sustainability certification process and, finally, to its consideration as a CORSIA Eligible Fuel. Let us begin by identifying the CORSIA Sustainability Criteria for Sustainable Aviation Fuels. The use of SAF expands as technology and research continue to develop in that direction. The ICAO Assembly acknowledged the need to develop and deploy SAF economically feasible, with processes that will not affect society or the environment. Sustainability Certification Schemes (SCS) have been implemented to verify economic operators' compliance with voluntary or mandatory sustainability criteria. The production and use of SAF must meet specific criteria set by the ICAO CORSIA framework. The first criterion is that SAF cannot be produced from biomass from high-carbon stock land converted after 2008. The second criterion is that SAF must be made through a life cycle process with at least a 10% emissions reduction. ICAO has developed the life cycle emission value for sustainable fuels to achieve and control this value. Additionally, SAF producers can calculate the life cycle emission values through the 'CORSIA Methodology for calculating actual life cycle emissions values'. These criteria will be confirmed by the International Carbon and Sustainability Certification (ICSC), which is an SCS already in action.[6]

*SAF Feedstock and Fuel Conversion*

Various sources for feedstock types could produce a CORSIA Eligible Fuel (CEF), many of those explained earlier in this chapter. The Committee of Aviation Environmental Protection (CAEP), in February 2019, developed default values for life cycle emission for CORSIA Sustainable Aviation Fuels produced from sixteen different types of feedstocks. Various methodologies are still under development for CORSIA lower-carbon aviation fuels. Since the CEF industry evolves, more feedstock types may become available to fuel producers. The feedstock types that generate aviation fuel go through approved fuel conversion processes. The International Standard-Setting Organization, ASTM International, has certified six fuel conversion processes for SAF and CEF. These certifications are relevant to the technical specifications of fuels and ensure that the outcome is safe for commercial use in aircraft by meeting the same safety standards as any other jet fuel. Then fuels must go through a sustainability certification process besides the technical certification process if they are used as CORSIA fuels. CAEP developed a sustainability certification process based on existing sustainability approaches, which can be either regulatory or voluntary, that prove that these are Sustainable Aviation Fuels.

*Life Cycle Emission Value (LSf)*

The use of CEF should decrease CO2 emissions in flights throughout their entire life cycle, meaning from the feedstock used, how it was cultivated and produced and the fuel conversion process followed until its combustion. All these factors combined create a fuel's life cycle emissions value (LSf). The CORSIA implementation element for CEF is a set of documents that include all the procedures and requirements for a CEF to be considered under CORSIA, all included in Annex 16 Volume IV. The explanation of these documents is briefly presented next:

1. CORSIA eligibility framework defines the SCS requirements that should be approved by ICAO to issue the sustainability certification of CORSIA Eligible Fuels and assess the life cycle emission value (LSf) of CEFs.

2. CORSIA-approved Sustainability Certification Schemes will include the list of SCSs that have been approved by the ICAO.
3. Sustainability criteria for eligible fuels present the sustainability criteria that need to be observed by a given fuel. The document's first edition applies until the end of the CORSIA pilot phase on December 31, 2023.
4. Default life cycle emissions values for CORSIA Eligible Fuels includes a list of default life cycle emissions values for CEFs, for the feedstock, conversion process and production region.
5. CORSIA methodology for calculating actual life cycle emissions values is a document that provides methodologies that fuel producers can use to calculate actual life cycle emissions values.

These methodologies allow fuel producers to claim life cycle emissions values lower than the default values. The Sustainability Certification Schemes (SCSs) will ensure that a CEF meets the CORSIA Sustainability Criteria (3). Then SCSs will ensure that the life cycle emissions value of the CEF is properly obtained (4 and 5). The ICAO council will approve the SCSs through this sustainability certification process (1), (2). Based on the previous five procedures, a methodology is developed to calculate carbon emissions reduction using CEF. The method in the next section describes how emissions are calculated and registered.

1. The operator obtains the CEF's life cycle emissions value (LSf). The CEF sustainability certification process determines this value.
2. The formula to calculate the CEF emissions reduction (ERy) is shown in the next section.
3. The operator includes information on CEF in its Emissions Report:

   - CEF emissions reductions (Ery) claimed;
   - fuel type, mass and life cycle emissions value (LSf); and
   - evidence of compliance with CORSIA Sustainability Criteria.

$$\mathrm{ER_y} = \mathrm{FCF} * \left[ \sum [MS_{f,y}] * \left( 1 - \frac{LS_f}{LC} \right) \right]$$

Ery = emissions reduction
FCF = fuel conversion factor, a fixed value 3.16 for jet A/Jet A1 or 3.10 (kg $CO_2$/kg fuel) for AvGas/Jet B
$MS_{f,y}$ = total mass of CEF claimed in the year y, by fuel type f [tonnes]
$LS_f$ = life cycle emissions value
LC = baseline life cycle emissions, fixed value, 89 for jet fuel or 95 for AvGas [g$CO_{2e}$/MJ].

Then this information is verified by a verification body on CEF provided in the Emissions Report. Then each state will collect and aggregate the verified information on CEF from all air operators. Then the state will report this information to ICAO through the CORSIA Central Registry (CCR).[5, 6]

### The Use of SAF and the Sustainable Development Goals

The production of SAF can develop new economic opportunities in farming communities, support environmental protection and improve aircraft environmental performance. In 2015, the United Nations announced the 2030 Agenda for Sustainable Development and set the 17 Sustainable Development Goals (SDGs). The 17 SDGs address the root causes of poverty and drive development for all sectors. Aviation, as mentioned in Chapter 1, supports fifteen of these goals. The increasing use of SAF helps to work towards SDG 7 (affordable and clean energy) and SDG 13 (climate action). Through the diversification of feedstock supply, the commercialisation of SAF can also help support some of the more socially and economically-focused SDGs, such as 'no poverty' and 'reduced inequalities', by providing employment opportunities in the Least Developed Countries. As the production of SAF is scaled up, the industry will also focus on avoiding negative impacts on SDG 6 (clean water and sanitation) and SDG 15 (life on land). SAF could contribute to a minimum of 75% fewer emissions than jet fuel from production, distribution, transportation and combustion. Simultaneously, the consumption of SAF emits less harmful emissions, such as particulates and sulphur, by 90% and 100%, respectively.

- **Refuel EU:** Refuel EU is an initiative by the European Union to accelerate the deployment of sustainable alternative fuels in the transportation sector. It is part of the EU's broader strategy to reduce greenhouse gas emissions in the transportation sector, which is accountable for about 25% of the EU's total emissions. The initiative aims to increase the use of low-carbon fuels, such as renewable energy, advanced biofuels and hydrogen in transport, as part of the EU's efforts to achieve its climate and energy targets for 2030 and become carbon neutral by 2050. The initiative aims to support alternative fuels and infrastructure across Europe, including electric vehicle charging stations, hydrogen refueling stations and advanced biofuel production facilities, seeking to develop a framework for the sustainable production of low-carbon fuels. When the European Union published the 'Fit for 55' legislative package in July 2021, the Refuel EU proposal was part of it, aiming to boost the production and use of SAF. The proposal mainly focuses on a blending approach for aviation fuel suppliers. Their obligation is that all airports in EU Member States can provide aircraft operators with a minimum share of SAF and a minimum share of synthetic fuel. This mandate is expected to begin in 2025, where the minimum share of SAF is 2% and with a five-year increase. The goal is to reach at least 63% of SAF by 2050, where 28% will be synthetic fuel.
- **The U.S. Renewable Fuel Standard (RFS):** It is a federal program established in 2005 that requires a specific volume of fuels, such as biofuels, to be blended into transportation fuels sold in the United States. The RFS is intended to reduce greenhouse gas emissions from the transportation sector, reduce dependence on foreign oil and support the development of a domestic biofuel industry. The renewable fuel standard (RFS) program was established by the U.S. Environmental Protection Agency (EPA) to promote renewable fuels in the transportation sector. The RFS requires a certain amount of renewable fuel to be blended into transportation fuel each year, aiming to increase the use of renewable fuels and reduce greenhouse gas emissions from the transportation sector. Under the RFS program, advanced biofuel use, including Sustainable Aviation Fuels (SAF), is incentivised. SAF is considered an advanced biofuel because it can reduce greenhouse gas emissions by up to 80% compared to conventional jet fuel. The EPA sets annual volume requirements for advanced biofuels, and obligated parties, such as refiners and importers of transportation fuel, must meet these requirements by using alternative fuels or purchasing credits called renewable identification numbers (RINs). SAF producers

can generate RINs for fuel use, which can be sold to obligated parties to meet their advanced biofuel volume requirements. This creates a market incentive for the production and use of SAF.

- **The California Low-Carbon Fuel Standard:** The policy was updated in 2019, including SAF as an eligible fuel to generate credits. This policy framework aims to put a value on carbon reduction produced from renewable sources. The GHG benefits from SAF are quantified from a Life Cycle Assessment process that compares the avoided emissions from conventional fuel. These credits incentivize the SAF generation, as they can be sold to other parties under the CA-LCFS.

Synthetic fuels, or e-fuels, are a significant development to contribute to the decarbonisation of the industry, with yet many withstanding challenges. Synthetic fuels are produced from a power-to-liquid (PtL) process. When the process is powered by renewable energy, it can result in a carbon-neutral flight emissions generation.

## Conclusion

This chapter explained the basic principles of Sustainable Aviation Fuels. The aviation sector is seeking opportunities for using less harmful aviation fuels. This means it is imperative to enhance the development of new alternatives. Fuels which are more environmentally friendly will have fewer emissions during aircraft flights and their production and with sustainable practices during their primary resources' extraction. Sustainable sources and natural feedstocks are the primary sources of producing Sustainable Aviation Fuels (SAF), the term used in aviation and broader to aviation biofuels. SAF is a blended fuel. The final SAF product is conventional kerosene with a fossil-based chemical structure blended with hydrocarbon and certified as a 'Jet A1' fuel. Sustainable Aviation Fuels can provide significant reductions in overall CO2 emissions, which can be up to 80% compared to fossil fuels. SAF can significantly reduce sulphur dioxide and other particulate matter emissions. However, SAF production is based on a low-carbon emission practice. The LCA of the feedstock production and the processes to make the fuels distinguish biofuels or biobased fuels from sustainable fuels. So far, eight production processes have been certified for blending with conventional aviation fuel, and still, more are pending approval.

Current SAF production technology allows sources such as municipal waste, used cooking oil and agricultural waste. In addition, crops and plants are also used for that exact purpose. The life cycle of SAF relies heavily on the processing of primary resources, meaning the crops and the way the crops are cultivated. The actual name of this type of fuel signifies that sustainability must be the core element to be met. ICAO CORSIA is the leading international scheme that supports promoting and using SAF in flights. To have fuels eligible to meet CORSIA requirements, they must meet certain criteria, with specific feedstock sources and a measurable life cycle emission value. The development and use of SAF is an excellent advancement for the aviation sector decarbonisation; however, there are still many challenges to overcome until it becomes fully available for commercial use.

**Chapter Review Questions**

*Figure 6.4* Airplane no. 6 pointing right

6.1 What are the main fuels used in aircraft flights, and what are some of their distinct characteristics?
6.2 Why is there a need to shift from conventional to unconventional fuels in the aviation sector?
6.3 What are the three main elements that constitute Sustainable Aviation Fuels?
6.4 What are some environmental benefits from Sustainable Aviation Fuels?
6.5 What is the difference between Sustainable Aviation Fuels and biobased fuels?
6.6 How do you explain the term 'life cycle emissions reduction', and why it is important for the Sustainable Aviation Fuels?
6.7 What is the role of the American Society for Testing and Materials (ASTM) International in the Sustainable Aviation Fuels production?
6.8 Explain the basic sources used to produce Sustainable Aviation Fuels?
6.9 Explain if mono-cropping is part of a sustainable process to produce Sustainable Aviation Fuels and why.
6.10 What is the pathway for conventional fuels' production and consumption?
6.11 What is the pathway for unconventional fuels' production and consumption?
6.12 What is a CORSIA Eligible Fuel?
6.13 What are the CORSIA Sustainability Criteria for SAF?
6.14 What is the process for approving a CORSIA Eligible Fuel?
6.15 How can SAF help meet the Sustainable Development Goals?

**Key Points to Remember**

- The phrase 'aviation fuels' caused a lot of controversy and arguments about aviation's position in environmental sustainability in the past decade.
- The necessity to seek less harmful aviation fuels leads to the development of new alternatives, more environmentally friendly, fewer emissions and sustainable practices of primary resources extraction and fuel production.
- Sustainable sources and natural feedstocks are the primary sources of producing Sustainable Aviation Fuels (SAF).
- Sustainable Aviation Fuels (SAF) is the term used by the aviation industry, given that this term is broader to aviation biofuels.
- The term 'biofuels' generally is used to explain fuels generated from biological resources, like plants or animal material.
- SAF is made of conventional kerosene, which has a fossil-based chemical structure blended with renewable hydrocarbon. It is certified as 'Jet A1' fuel and can be used without requiring technical modifications to be an engine or aircraft fuel systems.

- The method of creating sustainable and unconventional jet fuel is supported by producing fewer CO2 emissions in fuel consumption and production through its entire life cycle.
- Sustainable Aviation Fuels can provide significant reductions in overall CO2 emissions, which can be up to 80% compared to fossil fuels.
- Using SAF can significantly reduce sulphur, sulphur dioxide and other particulate matter emissions.
- The LCA of the feedstock production and the processes to make the fuels is basically what distinguishes biofuels or biobased fuels from sustainable fuels.
- There are eight production processes that have been certified for blending with conventional aviation fuel, and still, more are pending for approval.
- Current technology for SAF production allows sources such as municipal waste, used cooking oil and agricultural waste. In addition, crops and plants are also used for that same purpose.
- The life cycle of SAF relies heavily not only in the processing of primary resources, meaning the crops, but also from the way the crops are cultivated. The actual name of this type of fuels signifies that sustainability must be the core element to be met.
- CORSIA greatly supports the promotion and use of SAF in flights. To have fuels eligible to meet CORSIA requirements, they must meet certain criteria, with specific feedstock sources and a measurable life cycle emission value.
- 'Sustainable fuels' and 'alternative fuels' have been used interchangeably. They demonstrate produced fuels from non-conventional processes and resources with reduced environmental impacts.
- The feedstock types used to generate aviation fuel go through approved fuel conversion processes. The International Standard-Setting Organization, ASTM International, has certified six fuel conversion processes for SAF and CEF.
- The use of CEF should decrease CO2 emissions in flights throughout their whole life cycle, meaning from the feedstock used, how it was cultivated and produced and the fuel conversion process followed until its combustion.

## Acronyms

*Table 6.1* Acronym rundown

| | |
|---|---|
| SAF | Sustainable Aviation Fuels |
| AvGas | Aviation Gasoline |
| ASTM | American Society for Testing and Materials |
| FT-SPK | Fischer-Tropsch Synthetic Paraffinic Kerosene |
| HEFA | Hydroprocess Fatty Acid Esters and Free Fatty Acid |
| HDS-SIP | Hydroprocessing of Fermented Sugars – Synthetic Iso-Paraffinic Kerosene |
| ATJ-SPK | Alcohol-to-Jet Synthetic Paraffinic Kerosene |
| CHJ | Catalytic Hydrothermolysis Jet Fuel |
| MSW | Municipal Solid Waste |
| CEF | CORSIA Eligible Fuel |
| SCS | Sustainability Certification Scheme |
| ICSC | International Carbon and Sustainability Certification |
| CAEP | Committee of Aviation Environmental Protection |
| LSf | Life Cycle Emissions Value |
| Ery | Emissions Reduction |
| FCF | Fuel Conversion Factor |
| $MS_{fy}$ | Total mass of CEF claimed in the year y by fuel type f |
| CCR | CORSIA Central Registry |
| PtL | Power to Liquid |

## References

[1] Air Transport Association Group. (2017). *Beginners guide to SAF* (No. 3). https://aviationbenefits.org/media/166152/beginners-guide-to-safweb.pdf
[2] European Parliament. (2020). *Sustainable aviation fuels*. www.europarl.europa.eu/RegData/etudes/BRIE/2020/659361/EPRS_BRI(2020)659361_EN.pdf
[3] ASTM International. (2019). *D7566–19b, standard specification for aviation turbine fuel containing synthesized hydrocarbons*. www.astm.org
[4] International Coordinating Council of Aerospace Industries Associations. (2019). *Advancing technology opportunities to further reduce CO2 emissions*. ICAO. Protection/Documents/Environmental-Reports/2019/ENVReport2019_pg116–121.pdf
[5] ICAO Secretariat. (2019). *Environmental report 2019 aviation and environment*. ICAO. www.icao.int/environmental-protection/Documents/EnvironmentalReports/2019/ENVReport2019pg228-231.pdf
[6] Klepper, G., & Schmitz, N. (2019). *CORSIA SAF certification with ISCC – the international sustainability and carbon certification scheme*. ICAO. www.icao.int/environmental-protection/Documents/EnvironmentalReports/2019/ENVReport2019pg188-191.pdf
[7] IATA. (2022). *Fact sheet: EU and US policy approaches to advance SAF production*. IATA. www.iata.org/contentassets/d13875e9ed784f75bac90f000760e998/fact-sheet – us-and-eu-saf-policies.pdf

# 7 Sustainable Aviation Processes and Techniques

**Chapter Outcomes**

At the end of this chapter, you will be able to do the following:

- Identify the relationship of safety in aviation sustainability.
- Distinguish the role of safety in social and economic sustainability.
- Identify tools that can support sustainable aviation operations.
- Explain the Lean Six Sigma model and its role to sustainable aviation operations.
- Identify the basic elements of a root cause analysis.
- Apply the root cause analysis as a method for aviation and aerospace sustainability.

**Introduction**

When we hear the word operations, we can imagine a group working together on a project to find the best possible and efficient solution with the appropriate tools available. The aim is to achieve a desired outcome with no delays, no mistakes in the outcome, under specific quality standards and characteristics, as a signed contract has set them. A sustainable operation is a set of processes that will lead to a required and prearranged result, where the three aspects; people, profit and environment are equally addressed. Now let us add this set of processes that constitute sustainable operations within the aviation industry context. The scope of sustainable aviation operations embraces safety as part of social and economic sustainability to proper aerospace manufacturing conditions, flight operations and communication in air traffic control (ATC).

*Figure 7.1* Air traffic control tower

This Chapter explains the role of safety in aviation as part of sustainable operations, showing how safety contributes not only to aviation's economic sustainability but also that it is part to social sustainability. Similarly, the relative new concept of cybersecurity is a critical aspect to consider as part of sustainable aviation operations. Under any systemic view, techniques that can unravel problematic processes that critically affect the

DOI: 10.4324/9781003251231-7

operational outcome of an aviation system, are essential. Tools such as Six Sigma and Lean Management can contribute to sustainable operations if appropriately adapted and used in the operational nature of different aviation activities. This chapter presents Lean Management, Six Sigma and root cause analysis as tools and methodologies to identify risks and their root cause, aiming to optimise processes, increase efficiency and manage and mitigate risks. The overall objective is to build and maintain a resilient system without compromising the aviation safety and the product quality of the sector. This system will be ready to boost productivity, economic and operational efficiency and employees' trust. These elements are all parts of what we call a 'sustainable system'. Going back and addressing repeatedly same issues or leaving problematic areas unfixed are reasons for wasting money, compromising safety and trust and, eventually, having a defective and unsustainable system, unable to remain operative and profitable.

## Safety and Sustainability

Safety is one of the main principles – if not the only – that the aviation sector embraces throughout the history. Nowadays, the safety context and safety philosophies have expanded to almost every activity. The safety value often translates as a company's viability within the aviation ecosystem. This is particularly true since – i.e. we cannot talk about flight operations unless they follow specific safety standards. We cannot consider an aircraft airworthy unless all maintenance tasks and services are accomplished based on strict regulatory requirements and from properly certified staff. We cannot consider flight crew and cabin crew competency unless they have received the appropriate training to carry out their duties and responsibilities. Eventually, almost all activities in the commercial aviation are developed around strict and thorough safety standards. The aviation regulation is very detailed, and the aviation authorities, FAA, EASA, CASA or any National Civil Aviation Authorities (NCAA), are stringent when implementing safety regulations.

Aviation professionals undoubtedly claim that safety is the primary goal of their work and that no one and nothing should ever interfere and compromise the safety of their work outcome. It is a powerful argument and should not be challenged in any case. A safety management system (SMS) must be in place to achieve proper safety standards in any aviation organisation. The ICAO Safety Management Manual demonstrates the essential elements a safety management system (SMS) should have. ICAO, FAA and ESARR have similar definitions of safety and SMS, presented next.

- For the FAA, SMS is the formal, top-down, organisation-wide approach to managing safety risk and assuring the effectiveness of safety risk controls. It includes systematic procedures, practices and policies for managing safety risks.[1]
- Under the same scope, ICAO defines SMS as a systematic approach to managing safety, including the necessary management structures, accountabilities, policies and procedures.[2]
- According to the EUROCONTROL Safety Regulatory Requirements (ESARR), a safety management system is a systematic and explicit approach to defining the activities by which an organisation undertakes safety management to achieve acceptable or tolerable safety.[3]

The three definitions show that processes and procedures must be established systematically, embracing the systemic structure – a holistic approach that embraces the entire system and the ways it operates. This is a necessity since, for any management system to be effective, the people from the top management to the lower levels should be able to support it and follow its policies and guidelines without affecting the existing aviation regulatory requirements. At the same time, any aviation entity must have the proper organisational support for an SMS to function efficiently and with the desired maturity levels. This support will come first and foremost from the top management and leadership. Reinforcement from the leadership means that there is a holistic safety culture. A holistic safety culture will consider not only the safety outcomes of the product or service it offers as an aviation company but also the safety of its personnel while performing their duties. Occupational safety and health are also issues to consider in a technical environment. In that case, we talk about a company that values and respects its employees – the internal social element. We expect to see these elements in a sustainable working environment – supporting social sustainability internally and externally. In this section, we will focus only on the operational aspect of aviation safety. Chapter 9 will present the necessary elements for a management system to work appropriately and efficiently.

For any management system to perform well and achieve the set targets, it must have all the necessary resources to operate effectively; otherwise, the outcomes will not be at the desired level. The necessary resources can vary from qualified people in correct positions, professional development of employees, recurrent training for maintenance personnel and application of new technologies for work performance to in-depth training of the appropriate personnel who will use such new technologies. Supporting a technical solution and staffing is essential when discussing a functional and mature SMS. From that perspective, safety requires specific elements to be in place, both technical and organisational. These factors eventually constitute a socially sustainable working environment.

Even though aviation safety is a core value of the operations, more than having an SMS in place is needed to be a sustainable environment. The implementation of any practice reflects the culture that exists in a company. Hereafter, the lack of a holistic safety culture could lead to severe consequences not only at an operational level but also to regulation violations or unethical business behaviour. Hence, safety is a vital element of aviation operations. However, someone could claim that if any aviation company could be 100% accident free, no matter the cost is a sustainable organisation from a safety perspective. The holistic approach of sustainability mandates embracing the whole system and all its operations and activities. Safety is not the only element that makes an aviation company a sustainable working environment. Safety culture, safe operations and support of an SMS are all critical for social and economic sustainability. The central notion is that an aviation business must always seek balance among the sustainability principles considering the operational system and its interactions with safety, productivity and profitability and respect for employees, customers and the natural environment.

### *An Aviation Safety and Sustainability Model*

ICAO Safety Management Manual second edition (2008) presented James Reason's safety space model. The representation of the safety space model can be further expanded and aligned with the concept of aviation sustainability. An aviation business should be able to balance both production/profit and safety.

Safety management system (SMS) is an essential management tool for aviation operations, aiming to ensure that procedures, practices and policies are in place to manage and mitigate safety risks. Safety risk considerations take part in all business planning and decision-making to build the safest and most efficient pathway possible. An aviation company can determine the factors leading up to operational risks and hazards and identify proper measures to control and mitigate them by using an SMS. However, even if an aviation management system has fully embedded an accident-free culture, still, being completely risk-free is an unrealistic expectation.

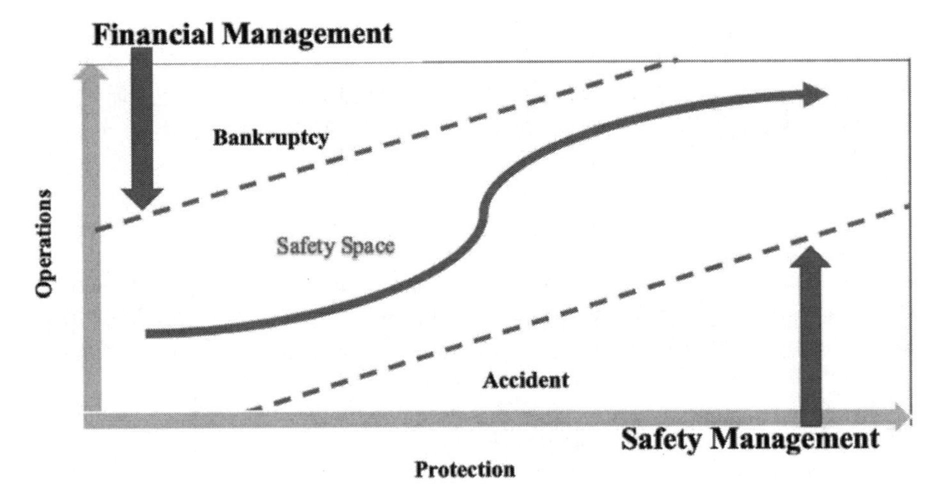


*Figure 7.2* The safety space[11]

An SMS is built upon four pillars: safety policy, risk management, assurance and promotion.

- Safety policy focuses on the policy and the organisation. It sets the objectives and standards, assigns responsibilities and gives upper management the platform to commit to a safety policy.
- Safety risk management focuses on risk mitigation, risk assessment and hazard identification and tracking. It will provide the decision-making process essential tools for reducing risks to an acceptable level through controls based on the organisation's capabilities and the operating environment particular needs.
- Safety assurance focuses on corrective action, internal evaluation programs and internal audits. It concentrates on continuously monitoring the safety performance of operational processes to ensure the controls in place are working as intended.
- Safety promotion focuses on culture, training and communication. It is designed to promote employees' 'solid foundation regarding their safety responsibilities, the safety policies and expectations, reporting procedures, and a familiarity with risk controls'.

The 'safety space', including protection and production, are related to safety and financial management systems in place, respectively. To achieve balanced operations, equal resources, support and monitoring must be provided to aviation operations' protection and production sides. Nevertheless, to maintain the 'safety space' between protection and production, an SMS needs to be designed with principles, focusing on sustainability's economic and social aspects. In that case, an organisation will require support for the longevity of its SMS.

First, an SMS will need to be financially sustainable. With proper assets to support the company's production and SMS integration and growth, the company will be economically viable in the long run. However, if it has economic prosperity without providing financial aid for SMS's needed resources, an incident or accident will not delay occurring. The sustainability aspect of an SMS should also rely on the existence and the correct allocation of resources. It is essential to provide economic support for the right reasons to the people, and this is also part of an economically sustainable organisation. Aviation operations require specifically qualified personnel, such as maintenance technicians and engineers and flight crew, but at the same time, personnel must get the assistance needed from their responsibilities. As mentioned in previous chapters, employees must follow recurrent training and professional development to enhance their working capabilities and utilise the defences against the human factor elements that will threat their work. This is an example of a combined economic and social sustainability approach. At the same time, when the company maintains a safety management system, personnel feel secure and respected by the leadership. In parallel, a satisfied employee will respect leadership and top management, enhancing their responsibility and motivation for their work. Therefore, the personnel that is satisfied and rewarded from their leadership will be more committed and responsible, being a vital part of preserving and safeguarding the safety management system. Hence, under a holistic and sustainable culture, personnel will be more involved in operations that fully support the company's viability.

Consequently, this approach becomes a continuous loop that returns to the same starting point – a safe and sustainable aviation system. An example of this approach is safety in aviation maintenance personnel and the role of sustainability. An MRO company can be characterised as sustainable when certain features exist. First and foremost, the MRO must have and maintain a safety management system under the relevant aviation regulation. The SMS must address the MRO's activities, services and organisational goals, with proper safety barrier in place. To maintain an SMS, a strong safety culture must be in place, embraced and supported by the leadership. Under that condition, employees will follow, support and embrace this culture. The safety culture, however, is a 'theoretical' concept without measurable features. A safety culture means that employees can work with the specific equipment, tools and technologies that will allow them to perform their duties on time, safely and under the organisational policies and the contracted agreements. This shows that the company can be economically viable since it will deliver its services and products. At the same time, human factors principles in aviation maintenance must be in place and consistently followed. The employees must work in a risk-free environment without factors compromising final product safety. This also means they must be physically and mentally capable. A sustainable aviation MRO should value these factors, offer the required assets to perform their work and eliminate the obvious or hidden risks. If an MRO does not possess a strong safety culture, inevitably, errors will occur. Employees will not trust their work environment, and this might cause implications for the final business outcome. Consequently, economic viability will be compromised as well.

## Cybersecurity and Aviation Sustainability

In the modern era, the aviation sector is pursuing the current technological trends and advancements in automation and digital information more vigorously than ever before. The aviation sector relies heavily on automation, data storage, cloud services and innovative digital information technologies, so it is often exposed to the threats behind these technologies. To mitigate these threats and their risks, cybersecurity systems and techniques are rapidly developing for aviation as for all other sectors. Cybersecurity is a series of processes to protect digital information and data from unauthorised access and use. It is essential to regard cybersecurity as an element of aviation sustainability since potential cyberattacks and their impact could significantly people any company's economic viability.

Cyberattacks refer to any malicious activities perpetrated by cybercriminals attempting to obtain, disrupt, degrade, destroy or otherwise compromise information systems or data, which can negatively impact employees, customers, other stakeholders and the aviation sector overall. As technology and connectivity expand, the opportunities for potential damages that could result from a cyberattack increase exponentially. Detecting a data breach can take up to six months, and this extended time frame prioritises prevention. Notably, the air transportation sector is the second most-targeted industry, following the financial sector. Some of the most significant threats in aviation include electronic flight bags (EFBs), in-flight entertainment connectivity systems (IFEC), NextGen air traffic controls (ATCs), onboard aircraft IP networks, aircraft interface devices, aircraft operation vulnerabilities and back-office tech platforms.[5] A cyberattack could procure steep regulatory fines, disrupt operations through extended delays or cancelled flights and damage the company's reputation, resulting in a loss of customer or stakeholder trust. In an extreme and worst-case scenario, a cyberattack could affect people's lives on a flight.

*Figure 7.3* Online privacy

For these reasons, a set of robust cybersecurity processes that could eventually be part of a cybersecurity system is an immense necessity. A cybersecurity system can include endpoint protection, cloud security, threat intelligence, identity and access management (IAM), cloud detonation centres, data protection and encryption, privacy by design, penetration testing, security training, awareness, competency and incident response – all factors that can prevent the consequences of cyberattacks, with proper processes and procedures.[6] The concept of cybersecurity can be a concise and more detailed approach consisting of specific and necessary methods and protocols to prevent, detect and eliminate the severe consequences of a cyberattack. Cybersecurity preventative measures are crucial because they are the first line of defence. To effectively control a cyberattack, a system must be proactive, maintain a detailed cybersecurity management system and be knowledgeable on global and industry-related threats.[7] Prevention could be best accomplished by complying with industry standards for information security. Detection is equally important because if a cyberattack is detected, it may be mitigated before the entire system is compromised. Proper detection measures could be applied by monitoring the network and information technology (IT) systems, protecting customer and operational data and acquiring knowledge on handling insider threats properly. The reaction is the final phase of a cybersecurity system, referring to incident response and mitigation actions. The measures of the reaction phase incorporate notifying customers and stakeholders of a breach, identifying security weaknesses and minimising the breach damage. The loop closes when new measures and processes are integrated to improve preventative measures.[8]

Another crucial area prone to cyberattacks is the expectation that the aviation industry will continue to rise in biometric authentication. Concerns have been voiced regarding the protection and preservation of biometric data. Interestingly, twenty-eight Member States in the European Union have regulations to protect biometric data.[9] However, in the United States, a comprehensive federal law still needs to cover the collection and use of biometric data. This is a significant cause of concern since a single cyberattack could leach a substantial amount of personal data, potentially jeopardising millions of people.

Additionally, considering the increased interconnectivity of the Internet of Things (IoT), it is necessary to have privacy protection measures in place to prevent or localise a breach of the sensitive data contained within these information systems.[8] Interpreting this information and its implications for aviation companies, it is realisable that protection from cyberthreats is immense, and cybersecurity could play a vital role in sustainability. As mentioned, it concerns the security and interests of the people while ensuring that the organisation, and on a larger scale, the sector, is sustainable for long-term success. Hence, a cybersecurity management system could be in place with the necessary resources and infrastructure to support all technological advancements.

The social element, people, must be able to identify and be protected from potential cybersecurity breaches. This familiarisation can be achieved through improving employees' competency with mentoring and supervision. Information security standards, best practices and regulations need to be adopted and implemented rigorously to protect the privacy of the stakeholders who trust that these organisations are secured against such threats.

Regarding the role of cybersecurity in the sustainability of an aviation environment, we could also recognise the role ethics, trust and values. Particularly, ethical concerns and trust issues can arise from passengers' data leaks. Those cyber risks might have impacts that can significantly affect any aviation company's economic status. Thus, it is necessary to ensure that the organisation has a well-prepared, detailed and inclusive cyber strategy to secure all company's assets.

Considering the sensitive nature of flight information systems and passenger data, ensuring that cyberthreats are managed holistically to protect against data exposures and support aviation's sustainability is critical. Systemic processes and defences must be developed to protect sensitive information. When passengers' credit card and personal details are not secured, this causes distrust in the company operations. Also, ethical implications arise when this data is misused to hack passengers' credit cards or use of those data for unlawful activities. Lack of trust, bad reputation and publicity will make passengers and employees uncertain and insecure. Therefore, internally and externally, the social element is directly affected. But how is the economic aspect affected? Well, suppose cyber breaches repeatedly happen in a system. In that case, eventually, passengers will not feel secure travelling not only of their data being leaked but also potentially for their safety as well. The company's economic indicators decline is a matter of time; therefore, it will not be viable anymore since the revenues will be reduced. Therefore, cybersecurity and sustainability go hand in hand when the company can provide the software, hardware and employee training for the appropriate and authorised use and protection of all digital information. Those prerequisites are supporting a cybersecurity approach and maintaining the system's economic and social sustainability.

## Lean Six Sigma

*Lean Six Sigma* is a tool that was not invented as a sustainability tool. It seeks to improve employee and company performance by eliminating the waste of resources and process/product defects. It combines the process improvement methods of Six Sigma and Lean methodology. The concept presented in this section is a set of tools that, if used appropriately in the correct working environment, can contribute to sustainable operations. Let us explore what Lean and Six Sigma mean and identify how they can contribute to sustainable company operations.

Lean and Six Sigma are two tools that complement each other. Lean accelerates Six Sigma, delivering more significant results than what would typically be achieved by Lean or Six Sigma individually. When combining these two methods, Lean Six Sigma can be a comprehensive tool that can increase the speed and effectiveness of any process of the value chain activity. Consequently, this will increase revenue, reduce costs and improve collaboration and work efficiency, contributing to economic and social sustainability.

> Lean methodology aims to optimise the people, resources, effort and energy of an organisation, creating, at the same time, value for the customer. It is based on two guiding tenets: continuous improvement and respect for people.

The Lean methodology involves a series of processes that constitute a management framework, Lean Management. The core philosophy of Lean Management is to reduce cost, eliminate waste, use resources appropriately and on time and deliver to the customers the value promised.

Lean is a method of streamlining a process, resulting in increased revenue, reduced costs and improved customer satisfaction. Mainly, in aerospace manufacturing, Lean processes are preferred because they are faster, more economical and efficient, and they deliver outputs with increased quality. A ubiquitous word often used in Lean methods is '*waste*' or '*removing waste*'. Under the Lean framework, '*waste*' is any unnecessary activity to complete a process. It should be noted that it is not the waste produced from a set of actions as a polluting factor. When the waste – the unnecessary process – is removed or corrected, only the required steps will be followed to produce the desired and promised customer service or product.

Summarising all the components needed to create a Lean manufacturing environment, someone can see it is not a matter of one or two things. It is the interaction and interrelation of several factors, as shown in Figure 7.4.

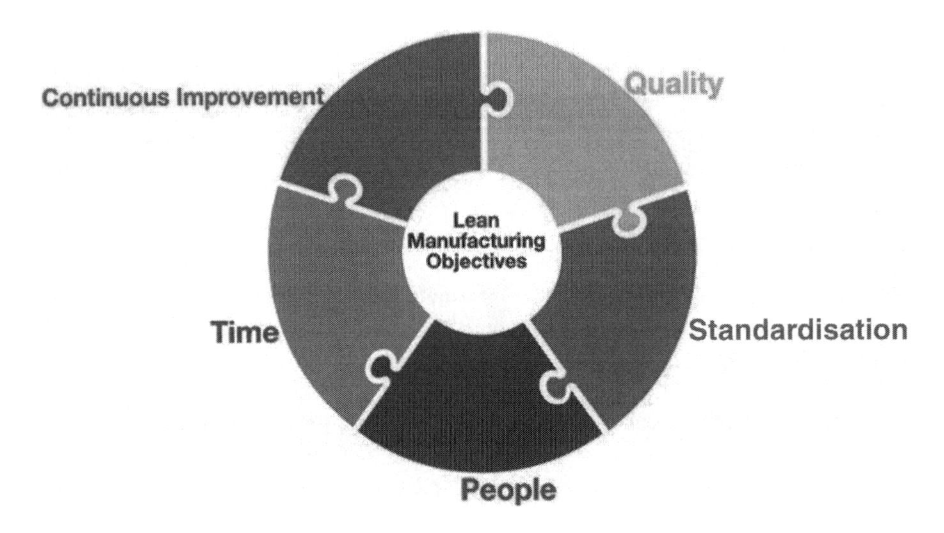


*Figure 7.4* Lean manufacturing factors

Moving on with the second element, Six Sigma is a method of efficiently solving a problem.

The name 'Six Sigma' refers to the statistical term 'sigma', which represents the standard deviation of a process. In the context of Six Sigma, it signifies the level of process performance. A higher sigma level indicates a more capable process and is less prone to defects. Six Sigma is a tool utilised for process improvement and quality management. Motorola originally developed it in the 1980s, and organisations across various industries have widely adopted it. Six Sigma uses a data-driven, evidence-based approach to identify and eliminate the causes of defects or errors, reduce process variability and improve overall process performance. It measures and analyses data to make informed decisions and implement effective solutions. Six Sigma is widely used in manufacturing, healthcare, finance, logistics and service sectors to achieve process excellence and operational efficiency. The Six Sigma methodology typically follows a structured problem-solving approach called DMAIC, which stands for define, measure, analyse, improve and control. Using Six Sigma reduces the number of defective products manufactured or services provided, resulting in increased revenue and greater customer satisfaction. Organisations utilise Six Sigma as a methodology that systematically and measurably enhances values, making them competitive, quality-conscious, customer-centric and forward-looking. Some of the benefits derived from the Six Sigma initiatives:

- waste prevention;
- defect reduction;
- cycle time reduction;
- cost savings; and
- market share improvement.

The primary key stakeholders and the focus of Six Sigma are the customers. This group of stakeholders is significant, and their needs should always be considered in a business context. They are also called *end user*. The final product or service must meet specific objectives and quality standards to satisfy the customer. This is an essential element to support economic sustainability. For an aerospace manufacturer, delivering an aircraft that does not meet the required design specifications will tremendously affect the manufacturer's financial viability. In an even more severe context, a faulty design could lead to incidents or fatal accidents that would compromise aviation safety. One way to avoid that is qualified personnel with specific backgrounds, knowledge and experience, leading us to the next important stakeholder of Six Sigma – the employees. Employees in the Six Sigma initiative must follow the necessary aerospace standards and design. Employees need support, efficient processes and ethical and healthy working conditions, with appropriate technical equipment. Finally, suppliers are the next stakeholder group, people who provide input to the process. Suppliers must respect and support internal processes. Hence, the company's operations are not compromised.[10]

### *The Benefits of using Lean Six Sigma*

The benefit of Lean Six Sigma is a methodology suggested to be applied under the sustainability lens, especially in production and design organisations. An increase in revenue through Lean Six Sigma is one of the most important benefits for a company. Economic sustainability relies heavily on preserving and distributing financial resources along the company's operations. Lean Six Sigma increases an organisation's revenue by correctly streamlining processes. Streamlining processes results in products or services being completed faster and more efficiently without compromising quality. For example, Lean Six Sigma could help increase revenue by enabling an aerospace company to do more with less: sell, manufacture and provide more products or services using fewer resources. This streamlining process leads to reduced costs.

Costly processes can be decreased when solving problems caused by or within a process. The issues are usually concerned with defects in a product or service that can increase administrative costs. That is why Lean Six Sigma enables one to develop and apply correct processes, which alternatively could cause an organisation to lose valuable resources.

Furthermore, Lean Six Sigma improves organisational efficiency by maximising efforts to deliver customers a satisfactory product or service. It allows for allocating resources and the revenue produced from newly improved processes towards growing the business. Moreover, Lean Six Sigma enables the creation of efficient techniques to deliver more products or services to more satisfied customers. In addition, Lean Six Sigma promote active participation and create engaged and responsible teams. Building trust is another benefit of the Lean Six Sigma methodology. Transparency throughout a business system promotes a shared understanding that each person contributes to success. Lean Six Sigma creates a purpose of ownership and accountability for employees, increasing their effectiveness at delivering results for any project they are involved. Regularly, this benefit is overlooked by organisations that implement Lean Six Sigma, but its underlying advantages dramatically increase the chances of continued business success.[10]

DMAIC model is based on common-sense practices and is completed in five phases:

- Define the problem and what is necessary to satisfy the customer.
- Measure the current process with appropriate data collection.
- Analyse through investigation of what causes the problem.
- Improve the process with proper solutions.
- Control: Sustain the improved actions.

The DMAIC model helps identify the cause of a problem and execute a solution based on facts rather than assumptions.

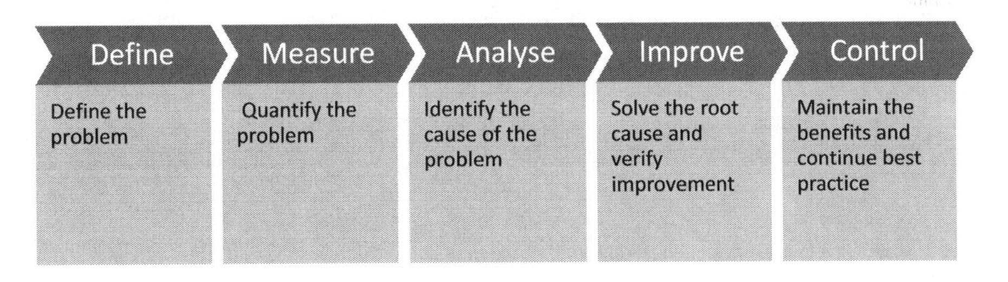


*Figure 7.5* The DMAIC steps

***Lean Methodology for Flights***

Six Sigma methodology includes individual sub-processes; it details every operation and the time of each one. It is a data-driven methodology for decision-making that could eliminate potential mistakes. As mentioned, the aviation sector implicates various interrelated processes with multiple challenges – applying the Six Sigma methodology in an airline where safety is the number one priority. The next is the continuous improvement of passenger satisfaction. However, due to the complexity of operations, there are many areas where things can go unexpected or even wrong, from delayed flights, missed flight connections, lost luggage, long layovers or cancelled flights. A combined approach of Lean Six Sigma could diminish or even mitigate these issues. The first step when applying the Lean Six Sigma methodology is to collect data and accurately determine the areas for improvement. The purpose is to identify wasteful processes. Processes involving passenger traffic flow, the number of delayed and cancelled flights and luggage loss is critical. At the same time, data can be gathered from customers' direct feedback using surveys and focus groups. One of the waste types often seen in airlines that apply the Lean Six Sigma methodology is wasted employee time.

Delay is another type of waste identified in Lean Management. Delayed departures affect numerous people every day. Six Sigma can reduce and eliminate many duplications, delays, redundant actions and misapplied rules that lead to these delays. If we translate these three factors into how they affect an airline's sustainability, finding ways to reduce wasted employee time is imperative. Working hours more than the designated does not mean people are productive. Employees' physical performance and ability to work under all safety and production requirements should be seriously valued. Proper time management of employees will make them productive and efficient, and the errors in work will likely be minimised, eventually forming an economically and socially sustainable working environment. Time is a critical factor for any aviation company, regardless of the final product or service they deliver. For a company to be and remain sustainable, it is necessary to eliminate, fix or improve all these processes that are a '*waste*'. Usually, processes causing waste are only sometimes obvious or transparent. As a latent condition, these processes continue to exist in a company's system, with uncontrolled consequences. This is what we call 'hidden factory'.

The hidden factory is the set of processes that can reduce quality or efficiency in a business's process and outcomes. It is often unknown to managers or others seeking to improve processes. Six Sigma focuses on identifying 'hidden factory' activities to eliminate their root causes. Six Sigma can contribute to identify hidden factory conditions, change status, improve profits and reduce wasteful processes. The hidden factory signifies the lack of sustainable processes and, thus, operations. When a system shows signs of reduced efficiency or low quality, it will eventually have increased costs, inefficient activities and consequences throughout the management system. Sustainability means that effective operations are in place, always striving to support the organisational system to reach excellence. A business strategy of an organisation encompasses a sustainability strategy too. If the root causes of a hidden factory are not appropriately defined, then strategies will not be adequately implemented, and consequently, their goals will not be met. Sustainability processes and Lean Six Sigma will mean identifying the sustainability principles through the management system.

***Lean Methodology for Maintenance, Repair and Overhaul***

All aircraft types and their components (i.e. avionics, engines, propellers, etc.) are subject to strict regulatory continuous airworthiness requirements, demanding regular maintenance. The maintenance services are performed by approved Maintenance, Repair and Overhaul (MRO) companies throughout the designated design life of the aircraft or aircraft part. The services conducted by an MRO are performed over various predetermined periods to provide life cycle maintenance. They include scheduled, out-of-service, preventative maintenance, repair, in-service modifications, rebuilding or retrofit. An MRO is a highly complex environment, bringing about many uncertainties while performing its services. These uncertainties appear in numerous forms, including unpredictable high demand, the scope of work, material requirements, inability to predict flow paths and limitations on technical data. As aircraft are under routine maintenance checks, additional issues may arise, which present other uncertainties for MRO timelines to complete the given tasks. Incorporating the Lean methodology can help respond to these circumstances and benefit the MRO and its customers. Lean is a way to optimize resources, effort and energy to create value for the customer and those performing the work; value is created by eliminating *waste*. By eliminating waste, an MRO can increase quality and flow, spend less time and effort on the tasks performed and lower the costs. Lean practices can support an MRO to assess its processes from the customer's viewpoint and what the customer considers valuable. It will aid in determining which methods provide value vs those that create waste and streamline the processes so there are no delays in the aircraft or part delivery. In addition, performing multiple tasks can be labour intensive. The location of these tasks is sometimes in different areas, requiring people to move around to accomplish the tasks, pursuing unnecessary whereabouts and time consumption. In aircraft maintenance, timelines for routine maintenance are often determined with certainty; however, challenges often arise when unforeseen work is discovered, such as cracks or corrosion. On top of that, unnecessary loss of time would compromise the MRO work program.

Additional challenges arise when an aircraft needs unscheduled maintenance, leading to shifting priorities and causing delays in the aircraft being repaired or overhauled. When multiple parts need to be worked on simultaneously, an employee may attempt to multitask. Multitasking can often lead to working on various projects or tasks simultaneously or moving to another one before the first is completed. This is one of the most significant delays in an MRO process. Waste elimination and value creation through the Lean methodology will allow an MRO to reduce changeover times, making the time needed to switch between products faster and respond more efficiently to customer demand. Lean also can help circumvent batch operations that create bottlenecks, queues, and increases in inventory. By avoiding batch operations, a flow is created.

Lean can also aid in organising equipment so that flow is an integral part of the MRO's processes, creating a value stream. If Lean methodology is appropriately implemented, MROs can determine the root cause of processes causing waste and respond more efficiently to customers' demands. The services must flow to reduce or eliminate long hauls. Lean also directly supports sustainability by reducing emissions, eliminating chemical use and protecting workers from injury. Through Lean, MROs can see an overall cost reduction, streamline their supply chain process, enhance their business processes and meet high-quality standards.

### *Sustainability Root Cause Analysis*

One expression mentioned several times in the previous paragraphs is the *root cause* of a faulty process. The root cause, as the words mean, explains the origin of a problem and allows identifying the source and fixing it. The root cause analysis can help resolve an issue in any aviation system and identify corrective steps. Eventually, correcting these faulty processes will improve the system's performance, by adopting more efficient practices, and enhance the company's economic sustainability.

The root cause analysis (RCA) is a method to address a problem or nonconformity and get to the 'root cause' of the problem. Its purpose is to correct or eliminate the cause and prevent recurring issues. Root cause analysis is a process of drilling down to find causes of the problem so corrective actions can be taken. Its focus is on systems and processes, not on individuals. A system with defective processes inevitably will not function properly.

#### *Why Is It Important to Determine the Root Cause?*

It would be very beneficial for a sustainable system to apply an RCA methodology since it will prevent problems from recurring and reduce the possibility of personnel injuries. It will also minimise rework and wasteful processes, increase competitiveness and, eventually, satisfy customers and stockholders. Ultimately, the correct application of RCA can reduce costs and save money for a company. Invariably, the root cause of a problem is not the initial reaction or response. It is not just restating the finding. The initial response is usually the symptom, not the root cause of the problem. Usually, the root cause of a problem is much more than just one faulty process. Especially when we talk about such a complex system as aviation, the flawed processes can be a set of processes or a program failure, a system failure, poorly written work procedures, lack of training or old equipment, among others.

A correct application of RCA will prevent the repetition of the same incidents. If the proper application is not achieved, resources will be continuously used. Employees will be occupied as a burden to other work tasks, causing them stress and untrust to the system they are part of, and consequently, services, products or any outcome will be compromised. Simply put, money will be lost, people will be less productive, customers will be unsatisfied and the organisation will be vulnerable to various risks. Reversing the stated, a proper systemic use of RCA could support economic and social sustainability in the aviation sector.

RCA applies after the immediate correction of a nonconformity. It is required when nonconformity is significant or could recur. An RCA methodology has a sequence of steps to follow, involving people, data collection and analysis and, most importantly, good system knowledge.

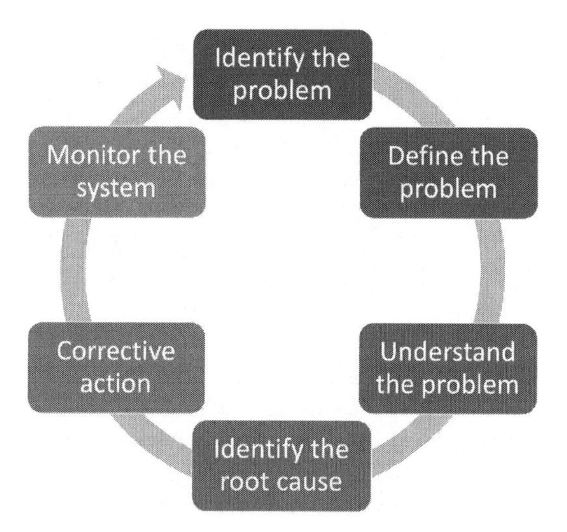


*Figure 7.6* Root cause analysis steps

- Assign the task to someone knowledgeable about the system and processes involved.
- Define the problem.
- Collect and analyse facts and data.
- Develop theories and possible causes – multiple interrelated causes may exist.
- Systematically reduce the possible theories and possible causes using the facts.
- Using different problem-solving methods to identify the underlying causes of problems or incidents.

**Questions to ask:**

- Who should work on the root cause analysis?
    - It is best if it is those who are closest to work daily.
- What's the problem? (define)
- Sometimes 'why' isn't the starting question.
    - What is it?
    - When did it happen?
    - Where did it happen?
    - How were overall goals affected?
    - Keep it simple – make an outline.
    - People see problems differently.
- Why did it happen? (analyse)
- What will be done? (prevent)

The fishbone diagram is a very popular tool that is used in RCA. The elements that constitute the fishbone diagram are as follows. The branches in the diagram can expand with more elements than those indicated in the next section, depending on the system, its complexity and size. With root cause techniques, someone can introduce detailed processes that will be specific to different areas within the system and support sustainable processes and practices.

- People (manpower) – training, verbal miscommunication, lack of communication, staff changed mid-project.
- Process/Methods – procedures, workflow, measurement (Calibrations performed, but were they appropriate?).
- Equipment/Technology (machines) – Defective, not maintained, not calibrated, overloaded.
- Management – annual reviews.
- Environment – Temperature, humidity, work area, distractions.
- Materials – Incorrect, degradation, certificates of analysis.

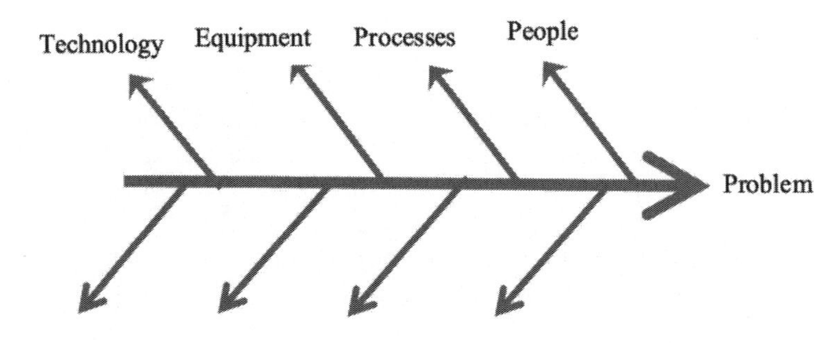


*Figure 7.7* The fishbone diagram

### Conclusion

While technology and network connectivity have played a substantial role in the sustainability of aviation and air travel through safer and more efficient operations, making these technologies secure is crucial. Aviation safety is a critical element for all aviation operations and is part of social and economic sustainability. In addition, in the last decade, governments and aviation organisations such as ICAO and IATA have emphasised aviation safety and cybersecurity. Cyberattacks can affect the day-to-day business operations of airlines, airport users' data and flights. All these can impact the economic and social sustainability of an organisation. As cyber awareness has increased within the aviation industry, international institutions and governments have begun to emphasise the importance of forming partnerships to develop a unified approach to cybersecurity. Airport managers and designers, for example, can establish cybersecurity policies and practices that can be applied worldwide. Professionals in avionics and electronic engineering can work together to develop a standardised solution for navigation and communication security risks. Manufacturers and engineers need to seek input from airlines and other operators to address potential threats as they arise. Collaboration between the various elements of the aviation industry is critical to minimise the hazards of cyberattacks on its businesses and customers to ensure the safety and sustainability of air travel. All aviation operations are highly interdependent and connected. Another methodology that supports sustainability operations in the aviation sector and enhances efficient processes without loss of time and money and without compromising customer satisfaction is Lean Six Sigma. The primary role of Lean Six Sigma is to eliminate wasteful processes that affect the system's proper operation. Root cause analysis is also presented, showing that the correct documentation of faulty processes and their problem identification will define the root of an issue, facilitating the implementation of a proper solution and eliminating that issue. Failure to do so will cost employee time and the company's money repeatedly spent on the same thing, causing potential problems with interrelated tasks and consequently affecting the company's sustainable operations.

### Key Points to Remember

- Safety is one of the main principles – if not the only – that the aviation and aerospace industry should follow and successfully meet. All aviation and aerospace activities are developed around strict and specific safety standards.
- The aviation regulation is very detailed, and the aviation authorities, FAA, EASA, CASA or any National Civil Aviation Authorities (NCAA), are stringent when implementing safety regulations.
- The three definitions show that processes and procedures must be established systematically, embracing the whole organisation structure. This is a necessity since, for any management system to be effective, the people from the top management to the lower organisational levels must support the system and follow its processes and procedures without deviating, in any case, from the aviation legislation.
- Support from the leadership means that the organisation holds a safety culture for the passengers, clients and employees. When an organisation cares about the safety of its personnel while performing its duties, we inevitably talk about a company that values its people.
- The social element is respected and protected. Under these conditions, the basic principles of ethics, respect and protection of a business environment are fully met and obtained. Without a doubt, it is a working environment that supports social sustainability internally.
- Any management system to perform well and achieve the set targets must have all the necessary resources to operate effectively; otherwise, the outcomes will not be the desired ones. Talking specifically about an SMS, technical and personnel support is essential; otherwise, safety standards are not met.

- Aviation safety is the core element of aviation operations. However, only having in place an SMS that is not properly working, then the outcome has severe consequences, and sometimes this might also mean regulation violation or unethical business behaviour. Hence, we must see aviation safety as a compulsory element of sustainable aviation operations.
- The 'safety space', including protection and production, has to do with safety and financial management systems in place, respectively. To avoid unbalanced operations, equal resources, support and monitoring must be provided to aviation operations' protection and production sides. Nevertheless, to maintain the 'safety space' between protection and production, an SMS needs to be designed with sustainability principles, focusing on sustainability's economic and social aspects.
- Another aspect of proper aviation operations that might compromise aviation sustainability is cybersecurity, if processes are not properly maintained. Cyberattacks refer to any malicious activities perpetrated by cybercriminals attempting to obtain, disrupt, degrade, destroy or otherwise compromise information systems or data, which can negatively impact employees, customers, other stakeholders and the aviation organisation itself.
- It is apparent that cybersecurity plays an important role in sustainability. It primarily concerns the security and interests of the people while also ensuring that the organisation, and on a larger scale, the sector, is sustainable for long-term success. While the aviation industry is not lacking in its regulations and procedures, the sector does not currently possess a cybersecurity framework that is tailored to its unique specifications.
- There is an intrinsic relationship between cybersecurity and sustainability within the aviation. Ethical concerns and trust issues will rise from passengers' data leaks in the case of a cyber breach.
- The economic and social impacts of cyber risks can affect significantly any aviation company; its sustainability relies on how well prepared this company is for such an event.
- *Lean Six Sigma* is a tool that was not invented as a sustainability tool. However, it can contribute to sustainable operations if used appropriately in a working environment. Let us explore what Lean and Six Sigma mean to identify how they can contribute to sustainable aviation operations.
- Lean accelerates Six Sigma, delivering more significant results than what would typically be achieved by Lean or Six Sigma individually. When combining these two methods, Lean Six Sigma can be a comprehensive tool that can increase the speed and effectiveness of any process within an organisation.
- The benefit of Lean Six Sigma is a methodology that must be applied for sustainability purposes, especially in production and design organisations.
- Sustainability in Lean Six Sigma will mean identifying the sustainability principles through the management system and its processes.

**Chapter Review Questions**

7.1 What is a safety management system, and why it is essential for all aviation operations?

7.2 What are the main pillars of a safety management system?

7.3 What is the role of safety in sustainable aviation processes?

7.4 Explain ICAO's safety space model and relate it to aviation sustainability.

7.5 What is cybersecurity, and what is its purpose for aviation?

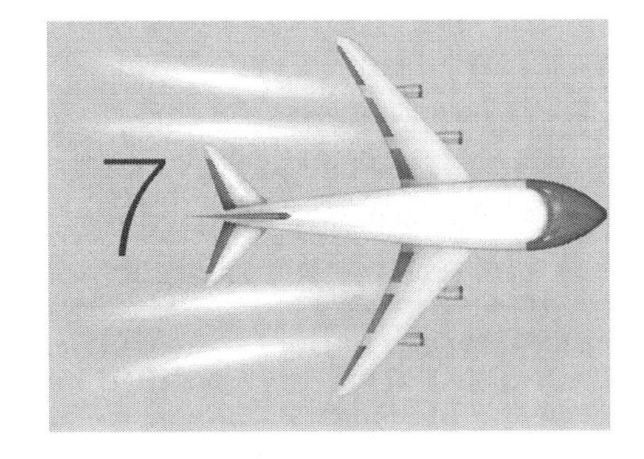


*Figure 7.8* Airplane no. 7 pointing right

7.6 How do you relate cybersecurity and aviation sustainability?
7.7 Explain how cybersecurity breaches could compromise aviation sustainability.
7.8 What are some vulnerable aviation operations to cyberthreats?
7.9 Explain the implications of cyberthreats to aviation safety for an airport.
7.10 How can cyberthreats affect air traffic control sustainable operations?
7.11 Explain the three main types of cyberattacks in an aircraft and ATC.
7.12 What does 'Lean' mean, and what is a Lean methodology?
7.13 How is '*waste*' defined in Lean Management?
7.14 How could Lean methodology and tools support sustainable processes of an aerospace manufacturer?
7.15 What is Six Sigma, and how it adds to an aviation company's sustainability and operations?
7.16 What are some benefits of using Lean Six Sigma, and how do they support an aviation company's sustainability?
7.17 What is a 'hidden factory', and how does it affect a company's operations and sustainability?
7.18 What is the root cause analysis?
7.19 Explain the role of root cause analysis in aviation sustainability systemic operations.

## Acronyms

*Table 7.1* Acronym rundown

| | |
|---|---|
| ATC | air traffic control |
| NCAA | National Civil Aviation Authorities |
| ICAO | International Civil Aviation Organisation |
| EASA | European Aviation Safety Agency |
| CASA | Civil Aviation Safety Authority |
| SMS | safety management system |
| DMAIC | define, measure, analyse, improve, control |
| RCA | Root Cause Analysis |
| EFBs | Electronic Flight Bags |
| IFEC | In-Flight Entertainment Connectivity Systems |
| IAM | Identify and Access Management |
| IT | information technology |
| ICT | Internet and Communication Technologies |
| NAS | National Airspace |
| ADS-B | Automatic Dependent Surveillance Broadcast |
| GPS | Global Positioning System |
| RCA | Root Cause Analysis |

## References

[1] Federal Aviation Administration. (2018). Safety management system: National policy. In *Federal aviation administration (8000.369B)*. U.S Department of Transportation. www.faa.gov/documentLibrary/media/Order/FAA_Order_8000.369B.pdf

[2] International Civil Aviation Organisation (ICAO) | *SKYbrary Aviation Safety*. (n.d.). Retrieved October 26, 2022, from www.skybrary.aero/articles/international-civil-aviation-organisation-icao

[3] *ESARR3 | SKYbrary Aviation Safety*. (n.d.). Retrieved October 26, 2022, from www.skybrary.aero/articles/esarr3

[4] ICAO. (2018). *ICAO SMS manual*. ICAO (Doc P859 version 4)

[5] Price Waterhouse Coopers. (2016). *Aviation perspectives 2016 special report series: Cybersecurity and the airline industry*. www.pwc.com/us/en/industrial-products/publications/assets/pwc-airline-industry-perspectives-cybersecurity.pdf

[6] Bose, R. (2019, October 23). *How can airlines protect their customers and data from evolving cyberthreats?* IBM Security Intelligence. https://securityintelligence.com/posts/how-can-airlines-protect-their-customers-and-data-from-evolving-cyberthreats/

[7] International Six Sigma Institute. (2013). *Six sigma – how six sigma DMAIC process work?* Retrieved January 27, 2023, from www.sixsigma-institute.org/How_Does_Six_Sigma_DMAIC_Process_Work.php

[8] Barajas, O. (2014, September 17). *How the internet of things (IoT) is changing the cybersecurity landscape*. IBM Security Intelligence. https://securityintelligence.com/how-the-internet-of-things-iot-is-changing-the-cybersecurity-landscape/

[9] Thales. (2021, June 16). *Biometric data and privacy laws (GDPR, CCPA/CPRA)*. www.thalesgroup.com/en/markets/digital-identity-and-security/government/biometrics/biometric-data

[10] International Six Sigma Institute. (n.d.). *Six sigma – what is the focus of six sigma?* Retrieved January 27, 2023, from www.sixsigma-institute.org/What_Is_The_Focus_Of_Six_Sigma.php

[11] GoLeanSixSigma.com. (2020, October 28). *DMAIC – the 5 phases of lean six sigma | GoLeanSixSigma.com*. https://goleansixsigma.com/dmaic-five-basic-phases-of-lean-six-sigma/

[12] International Six Sigma Institute. (2004). *Six sigma – what is the hidden factory*. Retrieved August 3, 2022, from www.sixsigma-institute.org/What_Is_The_Hidden_Factory.php

# 8 Sustainable Air Traffic Management

**Chapter Outcomes**

At the end of this chapter, you will be able to do the following:

- Define the basic principles of air traffic management.
- Explain basic air traffic management operations.
- Relate the air traffic management principles with sustainability.
- Identify operational threats in ATM and airports.
- Explain the role of Performance-Based Navigation in relation to environmental sustainability.

## Introduction

*Air traffic management (ATM)* relies on integrated, harmonised and internationally interoperable air transportation. As of ICAO, the ATM sector needs to go through changes at an operational level with a planning horizon beyond 2025. Air traffic in Europe before COVID-19, and for the last twenty years, is significantly growing. Still, few years after the COVID-19 pandemic, air transportation started to increase again since people began to fly again. However, even before COVID-19, the infrastructure in air traffic management for traffic flow monitoring has remained relatively the same over the past decades. Therefore, it must keep up with new technological developments and passenger needs. As of the European Commission, reliable and safe air transportation must always be maintained, increasing air capacity and use updated current air traffic control systems. ATM system must keep up with the fast economic development of aviation worldwide. Passengers need safe, efficient, affordable air transportation with fewer delays and better airport management. Respect and consideration for the environment are significant, mitigating risks compromising aviation safety, all necessary elements of a sustainable air transportation system.

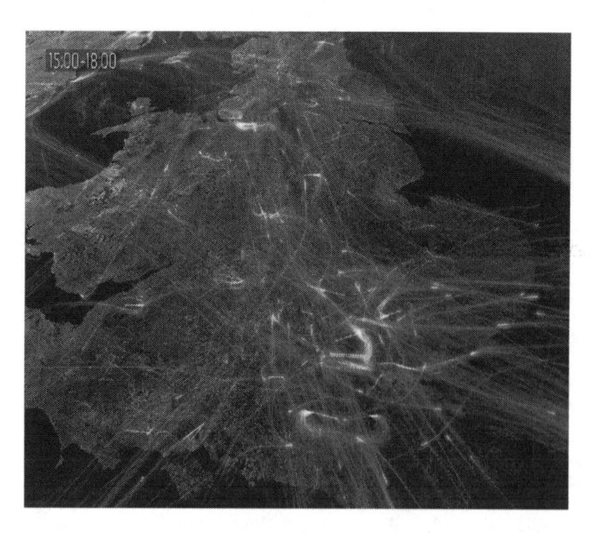

*Figure 8.1* Air traffic management network

During flights and air transportation, all users in an interoperable global air traffic management system must follow specific safety standards, provide and pursue

DOI: 10.4324/9781003251231-8

optimum economic operations, meet national security requirements and aim for sustainable practices. Let us have a moment and reflect on how these elements can lead to sustainable operations in air traffic management. Even indirectly, all three pillars of sustainability in an ATM environment can be met and, eventually, contribute to a sustainable aviation sector.

*Figure 8.2* Air traffic controllers

ATM is an integrated management system that encloses air traffic operations and airspace. It is a dynamic system supporting ATM operations safely, economically and efficiently, with clear and effective collaboration of all involved parties with improved facilities, new technologies and efficient operations. In addition, ATM is a system that requires coordinated operations of humans, with the appropriate information, necessary technology, facilities and services with air-, ground- and space-based communications, navigation and surveillance support. As part of ATM, air traffic control (ATC) is a vital element. ATC is a system to manage air traffic flow and ensure aircraft safety in the skies. ATC relies on human controllers, technology and procedures to guide aircraft from take-off to landing. As a complex system, ATC requires a high degree of expertise, skills and training. Controllers are trained to make quick decisions, communicate effectively with pilots and use advanced technology to manage the flow of air traffic. National authorities control the ATC system and ensure all aircraft comply with established rules and regulations. The primary role of ATC is to ensure that aircraft maintain a safe separation from each other and follow the correct flight paths. At the heart of the ATC system are radar and communication technologies. Radar systems track the position, speed and altitude of aircraft. Communication systems allow controllers to communicate with pilots during all stages of flight.

When an aircraft takes off, the tower controllers are responsible for managing the airspace around the airport. Once the aircraft reaches a certain altitude, communication passes to the approach controllers, who oversee the airspace around the approach to the airport. Finally, when the aircraft is close to the airport, the air traffic controllers are responsible for the aircraft's landing in certain runways under the traffic conditions of the airport and airways.

## Guiding ATM Principles and Sustainability

Air traffic management relies on several principles that define how an efficient ATM system operates. Like many other operations and techniques related to aviation activities, the ATM environment is also driven by strict safety standards and increased commercial and industrial expectations. There are explicit standards for global interoperability, and many States have systems that have evolved and still evolve within frameworks that can sustain their requirements. ATM relies on service provision and requires several resources, airspace, aircraft, humans and airports. ATM system controls flight between different airports through the airspace, safely and hazard free, within capacity limits, with optimum use of all system resources.

*Figure 8.3* Air traffic control radar screen

These features and their operations are based on the human factor involvement and capabilities, the whole infrastructure, technologies, interaction and interrelation of operations. However, as in all aviation operations, ATM needs support to meet the ever-growing global harmonisation and interoperability expectations. Sustainability in ATM will be applied and start from its principles: safety, humans, technology, collaboration continuity and information. Nevertheless, we must examine how sustainability could become part of these principles.

**Safety:** As with any aviation system, air traffic management requires high safety standards, with distinct processes in a structured and dynamic safety management system. ATM safety management system is necessary for efficient and effective services. As mentioned in previous chapters, safety in the aviation industry is part of social and economic sustainability. When safety standards are not adequately monitored nor maintained, or the required resources are not provided for air traffic management operations, the implications affect both social and economic sustainability.

**Humans:** Humans are central to the air traffic management system, as in all aviation operations. Humans are the process holders; again, the human factor element must be considered for errors, mistakes, omissions, etc. To minimise the factors that might compromise safety or economic conditions in the aviation sector, necessary resources, from current training to new technologies, hardware and software and related training, are essential in forming a sustainable working environment. Culture is also a valuable element for any human-centred system. Organisational culture can influence the system towards a sustainable culture, considering what needs to be done from people's verge.

*Figure 8.4* Airport terminal

**Technology:** Technology in air traffic management is critical for various operations. Navigation and communication systems, on the ground and air infrastructures, surveillance, data management and information technology are compulsory to reassure safe, efficient and effective communication with aircraft while on the ground or airborne. It requires fully integrated, interoperable and robust air traffic management across different regions with various traffic flows. Furthermore, people in this sector must have the skills and qualifications and remain updated on new technology advancements. A system to be sustainable must not only have the necessary people with the necessary skills to work in that environment to preserve all the safety standards but also have all the necessary technology available to preserve their working operations and comply with all the necessary safety standards.

**Information:** Air traffic management operations depend extensively on timely, relevant, quality-assured information to make informed decisions supporting air transport operations under high safety standards. This information must be shared promptly, efficiently and accurately, allowing all air traffic management community to conduct their business operations safely and efficiently. Communication is a critical element as far as it concerns a system's sustainability. We should not forget that aviation is a multicultural and international environment with people from different regions, countries and with different native languages. Communication and the information transmitted can be crucial elements for making a system operate sustainably; requesting clarification when information is not very well received and trying to use a more apparent accent or speaking slowly, especially when working in stressful conditions or being open to giving clarification or explanation to somebody who is not very fluent in the language used, are all factors that can make a system to operate under sustainability aspect.

**Collaboration:** Under the same concept of air traffic management information and communication, collaboration is a valuable factor characterising air traffic management operations. The people in air traffic management must have strategic and tactical collaboration. They all must participate and work in harmony to deliver all types and levels of service. Similarly, the air traffic management community must collaborate to maximise the system's efficiency by adequately sharing information with the appropriate quality assurance of data information to reach dynamic and flexible decisions for appropriate air traffic operations and management.

**Continuity:** All management systems must apply contingency measures when in high-risk operations. It includes specific processes that can mitigate risks, compromising air traffic safety, such as significant outages, natural disasters, civil unrest or any security threat, and other unusual circumstances. The continuity of operations constitute a viable and sustainable system.

Single European Sky ATM Research (SESAR) is the European air traffic control and infrastructure modernisation program. The program aims to develop the new generation air traffic management (ATM) system to ensure international air transport safety and fluidity for the next thirty years. The concept for the SESAR system is the EU's effort to address the air traffic management problem. The European Council established the SESAR project in 2005. SESAR is part of the Single European Sky, which lays down a clear organisation and establishes cross-border airspace blocks. These blocks change the status of ATM; routes and airspace structures are no longer defined by borders but by operational air traffic needs. SESAR has three stages of implementation. Given that the SESAR is for European operations, the implementation has several stages. This is due to the various air control systems in Europe and the varied nature of the fleet that is currently in service.

- The first phase, or the definition phase, was from 2005 to 2008. It included the basic air traffic modernisation plan, called SESAR ATM Master Plan, which set up different technological phases, schedules and prerogatives.
- The development phase from 2008 to 2013 included all possible essential technologies development that would underpin the new systems' generation.
- The third phase, the deployment phase, from 2014 and beyond, includes the large-scale installation of new systems and the widespread implementation of the related functions.

The Japan Aerospace Exploration Agency (JAXA) developed the Distributed and Revolutionary Efficient Air-Traffic Management System (DREAMS) project from 2012 to 2015. This project aimed to develop and integrate technologies that can contribute to a long-term vision for air traffic management to cover the upcoming global traffic growth. The DREAMS project has had notable benefits in the sector since new technologies were introduced. Weather information technologies expanded airport capacity and improved in-service rates with reduced effects from low-level wind disturbances and aircraft vortices. Noise abatement technologies maintained the noise exposure on the ground, with optimised approach routes to airports. High-accuracy satellite technologies improved the communication and navigation reliability between aircraft and on-ground services. Trajectory control technologies improved the in-service rates in airports where instrument approaches are not possible due to the geographical topography of the area. Trajectory control technologies enable a curved approach using global navigation systems (GNS) landing systems.

**Enhance Performance-Based Navigation**

The continued growth of air traffic and the need for flight efficiency mandates optimising the available airspace. Exploiting technological advancements in communication, navigation and surveillance enhanced air traffic management operations. Applying area navigation techniques in all flight phases contributes to improved airspace optimisation. Area navigation is enabled using an onboard navigation computer called an RNAV system. The capabilities of RNAV apply to maximise airspace capacity. Flight crew and ATC must apprehend RNAV system capabilities and confirm that these capabilities match the airspace requirements. RNAV systems' use lies at the core of Performance-Based Navigation (PBN).

Performance-Based Navigation (PBN) is a modern approach to air navigation that allows aircraft to fly more precise and efficient routes based on their performance capabilities. Rather than relying on traditional ground-based navigation aids, PBN uses advanced satellite-based technologies such as GPS and inertial navigation systems to guide aircraft.

The PBN system allows pilots to fly more direct routes, optimise their flight paths and reduce fuel consumption, reducing emissions and lowering costs. The system also improves safety by reducing the risk of midair collisions and allowing aircraft to fly in previously difficult areas to access due to terrain or weather conditions.

PBN is based on two key concepts: required navigation performance (RNP) and area navigation (RNAV). RNP measures the aircraft's ability to fly within a defined area with specific accuracy. At the same time, RNAV is a navigation method that allows aircraft to fly between waypoints using any suitable route.

Aircraft to use PBN must be equipped with navigation systems that meet specific performance standards and pilots with training and certification to operate the PBN system. Air traffic control must also be equipped with the necessary technology and trained to manage PBN routes and procedures.

The PBN concept determines three elements:

1. The navigation application is achieved using a NAVAID Infrastructure and associated navigation specifications.
2. The navigation aid (NAVAID) Infrastructure refers to the ground- and space-based navigation aids (except the non-directional beacon [NDB], which is excluded from use in PBN).

3. The navigation specification is a technical and operational feature that determines the navigation performance and functionality of an RNAV system. It also identifies the navigation equipment operations in the NAVAID Infrastructure to cover the operational needs identified in the Airspace Concept.

There are two kinds of navigation specifications: the RNAV and the RNP. Their difference is that an RNP specification requires onboard performance monitoring and alerting as part of the avionic functionality. The navigation specification provides material countries can use to develop their certification and operational approval documentation.

Performance-Based Navigation (PBN) redefines the aircraft's navigation capability from a sensor (equipment) based to performance based. The starting point of Performance-Based Navigation is area navigation – RNAV. The RNAV method permits the aircraft's operation on any desired flight path within the coverage of station-referenced navigation aids, within the limits of the capability of self-contained aids or a combination of these.

*Figure 8.5* Radar

With the start of area navigation, there was the requirement to define and standardize the navigation capability, which led to the development of a list of Performance-Based Navigation specifications. The aircraft must remain within a lateral value for each of these specifications to be approved. ICAO Member States, through their National Aviation Authorities, have many reasons and benefits for implementing PBN. First and foremost, aviation safety is the responsibility of all flights in their territories. Given the increased flight demand, PBN approaches, airspace and operation approvals will also continue to grow, requiring the involvement and acceptance of the NAA. The NAAs must fully know PBN to ensure a safe aviation operational environment for all its constituents.[4] PBN offers many benefits at an operational level, as shown in Figure 8.6.

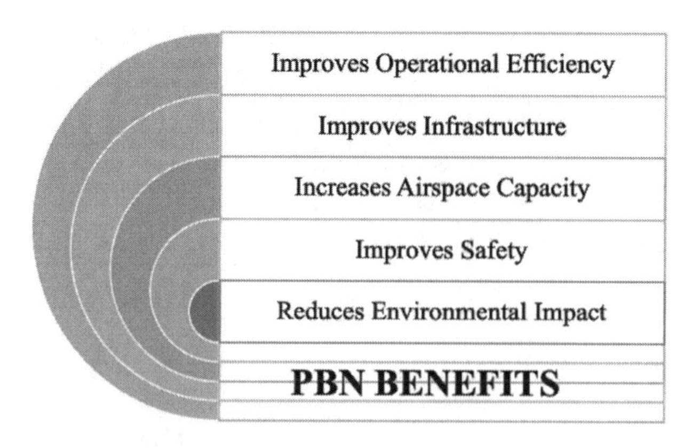


*Figure 8.6* Performance-based navigation benefits[4]

- PBN requires the use of an onboard RNAV system.
- PBN creates requirements for airworthiness certification and operational approval to use RNAV systems in airspace implementations.
- The RNAV system's functionality as well as its navigation accuracy in the NAVAID Infrastructure environment of the subject airspace must conform to the requirements stipulated in the relevant ICAO navigation specification.

For PBN, the aircraft and flight crew members must satisfy all the navigation specifications required during the flight in different airspace. From a controller's standpoint, PBN enables the systemisation of air traffic organisation through strategically deconflicting published ATS routes (including SIDs/STARs and instrument approach procedures) to reduce the need for tactical ATC intervention. PBN allows aircraft-to-aircraft separation to be 'built into' the airspace design, enabling migration from ATC to ATM.

***PBN and Environmental Sustainability***

The necessity for improved performance and functionality of navigation systems brought changes to aircraft design, separation minima – Reduced Vertical Separation Minima (RVSM), route spacing, airport access, procedure design and air traffic management. All these changes led to more efficient aircraft operations, reducing greenhouse gas emissions and aircraft noise levels. The PBN concept is an ICAO initiative developed in 2009. The requirement for performance, equipment functionality and improved infrastructure required the development of new navigation specifications. PBN has a practical implementation in flights, followed by documents, standards, regulations and certifications. Despite its original creation as a performance navigation tool, it enabled direct and indirect environmental benefits. First, fuel savings is one significant benefit. PBN offers options for sorter flight tracks, Optimum Profile Descent (OPD) capabilities, Continuous Climb Operations (CCO) and Continuous Descent Operations (CDO), also contributing to fuel savings in aircraft flights. In addition, PBN reduces flight variance through more predictable operations. It lowers the landing minima, reducing weather cancellations and diversions. One of the benefits of PBN in the fuel savings part is that it also reduces the probability of missed approaches or diversions and airport holdings due to airspace and airport capacity. Finally, PBN reduces the contingency fuel requirements owing to system predictability and reliability. Improving fuel savings leads to reduced CO2 emissions and other pollutants such as carbon monoxide or other nitrogen oxides.

Additionally, shorter flight paths and vertical profiles allow lower thrust levels with operations with lower noise levels. PBN also offers flight paths to non-noise-sensitive areas, with optimum drag profiles that support lower aerodynamic noise. Inevitably, the positive environmental effects are worthwhile. However, since through PBN operations, noise is also considered, residential areas or airport staff that are formerly disrupted by noise, the role of PBN in sustainability expands to the social element, except for environmental.

***Airport Terminal Cybersecurity***

Chapter 8 presents the role of cybersecurity as part of the techniques and processes that should be applied in an aviation system, focusing mainly on flight operations and maintenance organisations.

In this section, the role of cybersecurity is unravelled under the sustainability lens for air traffic management, focusing on airports and air traffic control.

Cyberthreats can be passive or active, and as we also discussed in the previous chapter, aviation operations must defend against any form of cyberthreat. Going deeper in our analysis for threats, passive threats involve only collecting data, whereas active threats require a transition between processes. Malicious abuse of open-source flight tracking information is an example of a passive threat that users must combat. These threats are typically less urgent than active ones but can still have severe outcomes, as they may go unnoticed for an extended time. On the other hand, active threats are easier to spot but could have more severe results. Large international airports could be compared to small cities due to their multiple and complex operations. Most of these operations are controlled electronically, storing data in cloud services. For example, some airports produce power through stand-alone renewable energy projects for their energy needs or recycle and reuse their waste. Modern airport terminals use 'smart' technologies to control and optimise their climate and lighting systems. Wireless internet is used for passenger convenience and everyday business airport operations. Ticketing kiosks, baggage check-in and tracking systems, computer terminals at each gate and point-of-sale terminals at the restaurants and shops inside the airport all utilise IP networks to conduct business. These are some of the few operations that can be disturbed by cyber risks.[5]

Outside infiltration of ticketing systems and airport security networks can expose passengers' personal information, including credit card and identification data. Malware, or malicious software, can be installed on airport computers and the computers of passengers who connect to public Wi-Fi networks, allowing hackers access to private data. Malware can even enter aircraft avionics systems while parked at terminal gates. Data breaches of airport users' data can be costly for airport operators, airlines and public opinion and trust. Consequently, the lack of cybersecurity in airport terminal networks can seriously affect the airport's economic and social sustainability. Thus, technology infrastructure is necessary to correct a violation and fortify airport systems to prevent repeated incidents.[6, 7]

### *Aircraft and Air Traffic Control Cybersecurity*

The increased connectivity of aircraft and air traffic control systems through the FAA's mandatory NextGen updates to the National Airspace (NAS) open them up to even more dangerous cybersecurity risks. One example is ADS-B (Automatic Dependent Surveillance-Broadcast), which monitors nearly every civil aircraft in the air to ease air traffic control and manage separation between aircraft. Pilots can see the precise locations of any nearby aircraft; air traffic controllers use this data to track aircraft and ensure no potential for spatial conflicts or collisions. One issue might be that since ADS-B operates on unencrypted IP networks and global positioning systems (GPS), it can be vulnerable to outside cyberattacks.

There are three types of attacks on ADS-B and GPS systems. The first type of attack is that GPS signals can be disrupted, making navigation more difficult for pilots and tracking more difficult for controllers. The second type of attack can obstruct communication systems and networks, making air-to-air communication between pilots and air-to-ground communications or between pilots and controllers more difficult or even impossible. The third type of attack is ADS-B communications, which can cause denial-of-service or modification or deletion of messages, causing confused communications between operators. Thus, aviation safety is highly compromised. Given the safety implications of missing or inaccurate navigation and locational data, the effect on sustainability must be addressed. Any attack on navigation and communication systems would result in reduced situational awareness by pilots and air traffic controllers. The worst possible result of a cyberattack would be a lethal aircraft accident. Apart from the losses of human lives, aircraft accidents typically bring on multiple lawsuits, additional regulatory scrutiny on the operator involved and public outcry for accountability, regardless of the operator's responsibility. Particularly for smaller airlines, this could have a disastrous effect on the operator's economic sustainability, even resulting in the forced buy-out or shutdown of the business. Social sustainability is also clearly impacted whenever there is a loss of life. There are other ways that cyberattacks can affect flight safety, but attacks on navigation and communication systems could be considered the most severe.[8]

The cybersecurity risks for aviation are not confined to airports and airborne aircraft. Commercial operators, cargo carriers, regulators and aircraft maintenance providers all utilize IP systems for booking, records and general business operations. Without proper cybersecurity measures, these organisations open themselves to attacks on internal financial data, maintenance records, customer data and more, which can shut down entire networks until they agree to pay a ransom. These scenarios can impact the economic sustainability of airports and air traffic control, also affecting airliners and related or involved stakeholders. Cybersecurity is necessary for the aviation sector since any system that has compromised safety and security operations automatically ceases to be sustainable.

## Conclusion

*Air traffic management* is a complex system requiring the interoperation of multiple systems, people and technologies. Making the aviation industry a sustainable system requires all its sectors to contribute to sustainability and become sustainable elements. Under that scope, this chapter identified some core activities that can contribute to the sustainability of the aviation sector. To bring change to a system, it is essential first to understand the basic principles that mandate its functions. Safety, humans, technology, information and continuity are the basic principles of air traffic operations. Each one of these elements must be maintained and supported so ATC operations are not compromised. The sustainability of ATC operations relies on preserving these elements so safety and operations are not compromised. Projects such as SESAR and DREAMS support the sector, offering enhanced navigation and trajectory services. The preservation and proper implementation of these technologies show the development of the sector and their contribution to sustainability. PBN is a tool describing an aircraft's capability to navigate using performance standards. However, its effects on environmental sustainability are notable in-flight performance, fuel consumption optimisation and emissions. Finally, the role of cybersecurity in ATM, navigation services and airport operations were presented in this chapter as part of cybersecurity and the elimination of cyber risks for sustainability.

## Key Points to Remember

- The ATM system relies on services' provision, is based on the provision of services and requires many different types of resources, airspace, aircraft, humans and airports.
- ATM needs help to meet the ever-growing global harmonisation and interoperability expectations.
- Sustainability in ATM will be applied and start from its principles: safety, humans, technology, collaboration, continuity and information.
- The concept for the SESAR system is the EU's effort to address the air traffic management problem.

- 'Single European Sky ATM Research' (SESAR) is the European air traffic control and infrastructure modernisation program.
- DREAMS project operated from 2012 to 2015. This project aimed to develop and integrate technologies that can contribute to a long-term vision for air traffic management to cover the upcoming global traffic growth.
- The PBN concept is an ICAO initiative developed in 2009. The requirement for performance, equipment functionality and improved infrastructure required the development of new navigation specifications.
- PBN has a practical implementation in flights, followed by documents, standards, regulations and certifications.
- PBN improves fuel savings and leads to reduced CO2 emissions and other pollutants, such as carbon monoxide or other nitrogen oxides.
- The lack of cybersecurity in airport terminal networks can seriously affect the airport's economic and social sustainability. Outside infiltration of ticketing systems and airport security networks can expose passengers' personal information, including credit card and identification data.
- The increased connectivity of aircraft and air traffic control systems through the FAA's mandatory NextGen updates to the National Airspace (NAS) open them up to even more dangerous cybersecurity risks.
- Any attack on navigation and communication systems would result in reduced situational awareness by pilots and air traffic controllers.
- Particularly for smaller airlines, this could have a disastrous effect on the operator's economic sustainability, even resulting in the forced buy-out or shutdown of the business. Social sustainability is also clearly impacted whenever there is a loss of life.

## Acronyms

*Table 8.1* Acronym rundown

| | |
|---|---|
| ICAO | International Civil Aviation Organization |
| ATM | air traffic management |
| ATC | air traffic control |
| PBN | Performance-Based Navigation |
| SESAR | Single European Sky ATM Research |
| DREAMS | Distributed and Revolutionarily Efficient Air-traffic Management System |
| JAXA | Japan Aerospace Exploration Agency |
| GNS | global navigation system |
| RNAV | area navigation |
| NAVAID | navigation aid |
| NAA | National Aviation Authorities |
| NDB | Non- Directional Beacon |
| RVSM | Reduced Vertical Separation Minima |
| OPD | Optimum Profile Descent |
| CCO | Continuous Climb Operations |
| CDO | Continuous Descent Operations |
| NAS | National Airspace |
| ADS-B | Automatic Dependent Surveillance-Broadcast |
| GPS | Global Positioning System |

**Chapter Review Questions**

8.1 What are the basic elements that can make air transportation to operate under sustainability principles?
8.2 What is the role of correct and effective air traffic management operations for supporting aviation sustainability?
8.3 How do the air traffic management principles correlate with aviation sustainability?
8.4 How do you define sustainable processes in an air traffic control?
8.5 How can each of ICAO's air traffic management principles add to sustainable aviation operations?
8.6 What is the role of communication in sustainable air traffic management operations?
8.7 What is the SESAR program?
8.8 How does SESAR program contribute to aviation sustainability?
8.9 What is the DREAMS project?
8.10 Are there any points in the DREAMS project that can be considered as sustainable operations?
8.11 What is the Performance-Based Navigation (PBN)?
8.12 What are the benefits of PBN operations?
8.13 Explain the role of PBN to aviation's environmental sustainability.

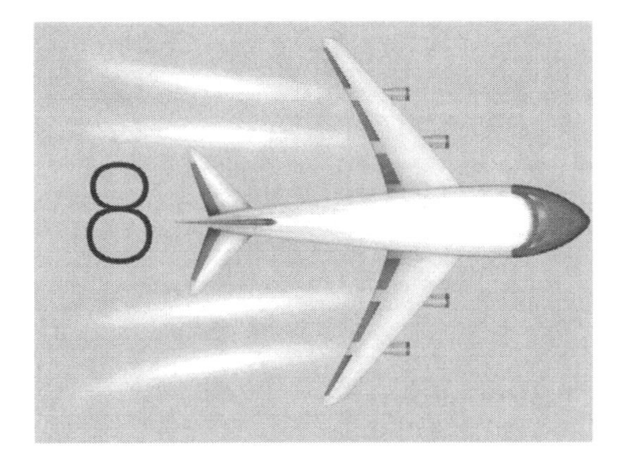


*Figure 8.7* Airplane no. 8 pointing right

**References**

[1] International Civil Aviation Organisation. (2005). *Global air traffic management operational concept*. www.icao.int/airnavigation/IMP/Documents/Doc%209854%20-%20Global%20ATM%20Operational%20Concept.pdf
[2] *SESAR | SKYbrary Aviation Safety*. (n.d.). https://skybrary.aero/articles/sesar
[3] Aviation Technology Directorate. (n.d.). *Distributed and revolutionarily efficient air-traffic management system (DREAMS) | Sky4All | Research & development | Aviation technology directorate*. www.aero.jaxa.jp/eng/research/star/dreams/
[4] ICAO. (n.d.). *Performance-based navigation (PBN)*. ICAO PBN. Retrieved December 10, 2022, from www.icao.int/safety/pbn/miscellaneous%20items/pbn%20faqs.pdf
[5] Hauser, C. (2020, May 19). *EasyJet says cyberattack stole data of 9 million customers*. www.nytimes.com/2020/05/19/business/easyjet-hacked.html
[6] FAA. (2019). *FAA has made progress but additional actions remain to implement congressionally mandated cyber initiatives*. U.S. Department of Transportation. www.oig.dot.gov/sites/default/files/FAA%20Cybersecurity%20Program%20Final%20Report%5E03.20.19.pdf
[7] Lykou, G., Anagnostopoulou, A., & Gritzalis, D. (2019). Smart airport cybersecurity: Threat mitigation and cyber resilience controls. *Sensors*, *19*(1). https://doi.org/10.3390/s19010019
[8] Haass, J., Sampigethaya, R., & Capezzuto, V. (2016). Aviation and cybersecurity: Opportunities for applied research. *TR News* (304). https://commons.erau.edu/publication/299

# 9 Sustainable Aviation Facilities

**Chapter Outcomes**

At the end of this chapter, you will be able to do the following:

- Define the elements that make sustainable aviation facilities.
- Identify the schemes appropriate for different aviation facilities.
- Explain the role of ISOs for creating a sustainable working environment.
- Learn the basic components of energy, environment and OH&S management systems.
- Explain the role of LEED for creating sustainable buildings for aviation.
- Identify the different types of aviation wastes.
- Describe the necessary components for a waste management system in airports.

**Introduction**

As explained, sustainability in aviation concerns various facets of operations, services and products. As we reflect on the term and as we have explored in previous chapters, the fundamental principle of sustainability is to apply to the whole system its products and services. The holistic sustainability approach is a unique concept that can change aviation and aerospace operations in a more viable and resilient pathway. Sustainability is not only about how the aviation industry works; it also concerns the buildings and facilities hosted by aviation businesses. How can we talk about sustainability in aviation and aerospace if the building that's an airline has its offices and has high energy consumption without considering energy efficiency measurements? Or how sustainable is a hangar if there are no appropriate ventilation systems; hence, employees are exposed to toxic and dangerous fumes? Or airports without recycling and proper waste management plans? Sustainable aviation needs facilities that will respect the environment, protect its employees and reduce energy costs and losses.

*Figure 9.1* Taipei Taoyuan International Airport[1]

It is interesting to mention that airline facilities and offices have started to get certifications under ISO 14001 and 50001. Airports can get certifications for their terminal

DOI: 10.4324/9781003251231-9

building with ISOs, LEED, BREEAM or any other environmental or energy management system. Leadership in Energy and Environmental Design (LEED) and Building Research Establishment Environmental Assessment Method (BREEAM) certifications; ISO management systems for environment, energy and Occupational Health and Safety; and waste and water management systems are the core topics of this chapter as an effort to identify the necessary components for making sustainable aviation facilities. Sustainability must be inclusive to aviation facilities as well not only in terms of reducing pollution and environmental effects and minimising waste but to also offer the best working conditions to employees and try to minimise costs from sensible use and consumption of all types of resources. Apart from covering these topics, the systemic theory behind all management systems is also presented, providing information on their implementation.

## Leadership in Energy and Environmental Design (LEED)

Leadership in Energy and Environmental Design (LEED) is a globally recognised certification of leadership and sustainability achievement in efficient, healthy, carbon and cost-saving green buildings. LEED is a scheme that addresses the triple bottom line of sustainability – people, planet and profit – by looking at the system holistically and considering factors in all building's critical elements: water, health and energy. LEED buildings use 23% less energy and 28% less water, significantly reducing fossil fuel consumption and energy costs. A LEED-certified structure also encourages users to compost and reduce landfill waste by shares up to 9%. This type of building also reduces the carbon footprint by up to 50%, using renewable energy resources, photovoltaic panels in the building, rooftop or an on-site electricity generation plant. LEED certification can benefit facilities with multiple operations and increased energy consumption, such as airports or airliner offices. Additionally, a LEED-certified building can offer improved indoor air quality suitable for travellers, employees and guests through ventilation and air conditioning systems with antimicrobial/anti-allergy filters. Inevitably, healthy conditions in a building lead to improved workforce productivity. With the public being more concerned about global warming and climate change, businesses are looking for ways to act more environmentally friendly. Furthermore, looking at the economic aspect of a LEED-certified building with multiple operations, such as an airport or offices, there is a significant reduction in operational and energy costs.[2]

For LEED certification, a project earns points by meeting specific requirements and earns credits that address anything that can address carbon emissions reduction, waste and water reduction and use, healthy and efficient materials and indoor air quality, among others. The projects go through a verification and review process from the Green Buildings Council, Inc., and points that correspond to a level of LEED certification are awarded. The stages are four:

- Certified (40–49 points earned)
- Silver (50–59 points earned)
- Gold (60–79 points earned)
- Platinum (80+ points earned)

*Figure 9.2* LEED sign[3]

Overall, buildings with LEED certification can save money, improve energy efficiency, reduce carbon emissions and create healthier facilities.

### BRE'S Environmental Assessment Method (BREEAM)

BREEAM is the world's first and foremost sustainability standard and rating system for the built environment launched in 1990. On a global level, over 540,000 buildings have certified BREEAM assessment ratings and more than two million are registered for assessment. The BRE's Environmental Assessment Method (BREEAM) is a voluntary scheme that rewards those designs that take positive steps to minimise their environmental impacts. It sets the standard for best practices in sustainable building design, construction and operation. It has become one of the most comprehensive and widely recognised measures of a building's environmental performance. For health reasons and improved work productivity, improving environmental quality and conditions is essential for all employees. It is vital to provide proper working conditions to people through healthy buildings and structures. Buildings with a lot of petroleum-based structural materials, without adequate ventilation, and unevenly increased temperatures on higher floors can cause adverse health symptoms to employees, such as discomfort, irritation, headaches and nausea. In more extreme cases, these conditions can even cause respiratory problems. These buildings are causing the known 'sick building syndrome'. For an aviation working environment, structures with conditions prone to generate all of the aforementioned are not only affecting working performance and productivity but they can also hinder risks as latent factors for minor incidents, errors, to severe accidents.[2]

> Building-related illness (BRI) concerns the often called 'sick buildings'. It means buildings constructed with toxic materials could cause allergies, asthma or other serious respiratory illnesses. This type of building does not usually have good and proper ventilation, leading to increased concentration of carbon monoxide from the construction materials. Buildings that also use a lot of plastic materials increase the effects of building-related illnesses.

### ISO Management Systems

Before starting the presentation of ISO management systems, a few basic details are offered to build the foundation for their proper implementation. The systemic approach of these standards requires the definition of some important terms, as shown in the following sections. ISO certification is not mandatory. However, any company or organisation that is ISO certified must follow the requirements to maintain the certification.

#### *What Is a Management System?*

A management system is a series of defined, organisation-wide processes that support practical decision-making–related actions. The decision-making actions aim to meet specific management system goals, such as quality goals, safety goals, environmental goals, etc. Hence, we have different types of management systems: quality management systems (QMS), safety management systems (SMS), environmental management systems (EMS), etc. The aviation legislation mandates quality management systems, safety and risk management systems and security management systems for the aviation sector. However, apart from the legislation, we have the ISO standards for companies if they wish to get certified. Nevertheless, when a company chooses to get certified, it must abide by that ISO standard's requirements, structure and maintenance. In both cases, an aviation management system or an ISO management system, there are two standard components to consider: the processes and the procedures.

*Processes and Procedures*

*Process* is a set of interrelated or interacting activities which transforms inputs into outputs. Inputs to a process are generally outputs of other processes. Processes in an entity are planned and carried out under controlled conditions to add value to its operations and final products or outcomes. A process where the conformity of the resulting output cannot be readily or economically verified is a 'special process'.

*Procedure* is the specified way to carry out an activity or a process. The procedures can be documented or not. The term 'written procedure' or 'documented procedure' is frequently used when a procedure is documented. The document that contains a procedure can be called a 'procedure document'. As previously mentioned, any activity, or set of activities, that uses resources to transform inputs into outputs can be considered a process. For organisations to function effectively, they must identify and manage numerous interrelated and interacting processes. Often, the result from one process will directly form the input of the following process. The systematic identification and management of the processes employed within an organisation, particularly the interactions between such processes, is called the process approach.[4]

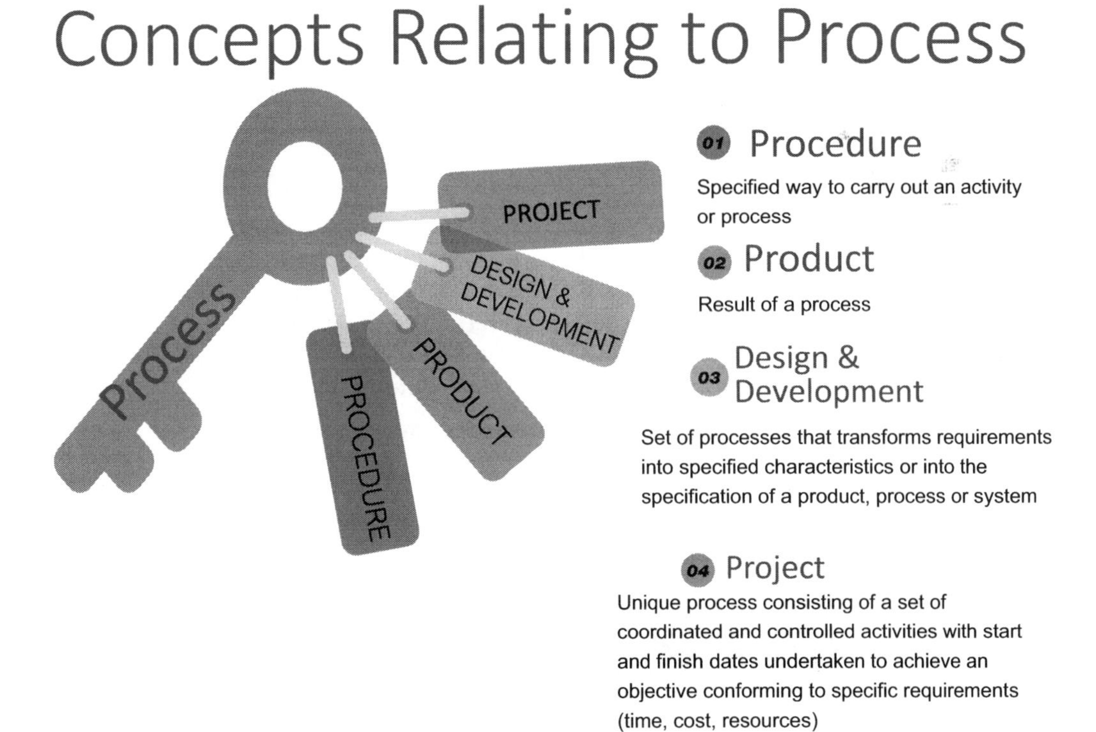


*Figure 9.3* The process approach

The process approach requires proper guidelines for developing a specific product. The design and development are based on the agreed outcome, and all design and development stages are defined. Finally, we have an accomplished project when all these steps are followed and completed. All the steps from the beginning to the end of the project are processes that transform inputs into specific outputs.

***ISO 14001 Environmental Management Systems***

ISO 14001 is an international standard for environmental management systems created by the International Organization for Standardization. ISO 14001 was first published in 1996 and updated in 2004. This third edition was published in September 2015, replacing all previous editions. This standard aims to help organisations protect the environment and respond to changing environmental conditions. According to ISO 14001, any organisation can achieve these goals if it establishes an environmental management system (EMS) and continually tries to improve its suitability, adequacy and effectiveness. ISO 14001 2015 is an environmental management standard. It defines a set of environmental management requirements. ISO 14001 includes a set of paragraphs providing specific requirements to be met, such as organisational context, leadership, planning, support, operation, performance evaluation and improvement. As with all management systems, ISO 14001 follows PDCA – plan-do-check-act model.[5] The PDCA model is a process improvement model that aims to support the proper monitoring and control of a management system. The ISO 14001 scheme contributes to sustainability in an organisation by creating an effective environmental management system that benefits the community and assists in meeting environmental regulations. According to the International Organization for Standardization, the ISO 14001 model provides a competitive edge through financial efficiencies, increased stakeholder confidence, strategic business aims, encouragement for environmental improvement of suppliers, community reputation and employee engagement. Overall, implementing ISO 14001 benefits the organisation and the surrounding community.

ISO 14001 applies to all organisation types. It does not matter what they do or what size they are. It can help any organisation to protect the environment and to respond to changing environmental conditions. According to ISO 14001 2015, an organisation's EMS must meet every requirement to comply with this standard. However, how ISO requirements are fulfilled will depend on many factors. It will depend on the organisation's context, structure, activities, objectives, compliance obligations and products and services and will be influenced by its risks, opportunities, environmental aspects and impacts. Once the EMS is established, it is used to do the following:

- Meet compliance obligations.
- Achieve environmental objectives.
- Enhance environmental performance.
- Facilitate and support sustainable development.

According to ISO 14001, any organisation can achieve these goals by establishing an EMS and then using it to manage the environmental aspects and impacts of its activities, processes, products, services and systems.

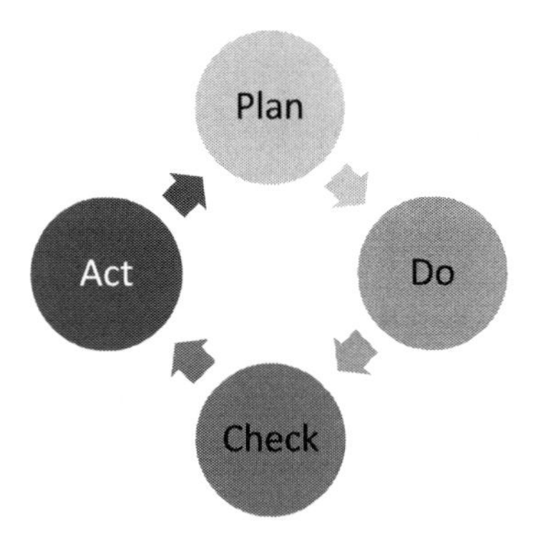


- The Plan-Do-Check-Act cycle is a four-step model for carrying out change in a management system.
- The change can be a new process entering the system or a change to an existing one.
- Just as a circle has no end, the model can be repeated continuously, aiming always to improve the system.
- The PDCA model applies to all management systems.

*Figure 9.4* The PDCA model: the continuous improvement tool

According to ISO 14001, an EMS will enhance an organisation's environmental performance since it will do the following:

- Reduce its environmental risks.
- Implement its environmental policy.
- Achieve its environmental objectives.
- Meet its environmental compliance obligations.
- Take advantage of its environmental opportunities.

An EMS can support sustainability in the aviation sector since it will show how to approach environmental management systematically. Such an approach will contribute to the organisation's long-term success and support sustainable development. It will do all of this by helping a company with the following:

- Prevent or reduce adverse environmental impacts.
- Implement environmentally sound practices and programs.
- Coordinate environmental initiatives with interested parties.
- Control how products are managed throughout their life cycle.
- Mitigate the adverse impact that environmental threats can have.

Once the EMS is established, meets ISO's requirements and deals with the organisation's risks and needs, it will ask the certification body to audit the system. If it passes the audit, the registrar will issue an official certificate that states that the organisation's EMS meets ISO requirements.[5]

***ISO 50001 Energy Management Systems***

The United Nations Industrial Development Organization (UNIDO) recognised that industries worldwide need effective action against climate change.

Energy is the major contributor to climate change, making up nearly 60% of the globe's GHG emissions; almost one billion people lack electricity access, and many rely on harmful polluting energy sources. ISO 50001 2018 is the latest standard for implementing a robust energy management system (EnMS), with the newest version published in 2018. It establishes an international framework for industrial plants or companies to manage all aspects of energy, including procurement and use. It assists organisations in making better use of their existing energy-consuming assets. As of the United Nations Industrial Development Organization (UNIDO), industries worldwide need to take action against climate change. It also noted that the modern world (China, Denmark, Ireland, Japan, the Republic of Korea, Netherlands, Sweden, Thailand, the U.S.A. and the European Union) must adopt national energy management standards to respond to market demand for energy efficiency. This standard provides a flexible road map for creating a certified energy management system that provides benefits, such as reducing energy consumption and energy costs.[6] Internationally, companies using the ISO 50001 during their early years reported between 5% and 30% savings in energy costs.[7]

ISO 50001 includes ten paragraphs, as all ISO standards; the paragraphs have the requirements to meet, starting from paragraph four, the organisation's context. Paragraph five describes the criteria for leadership, commitment and energy policy. Paragraph six sets the requirements for planning, energy risks, energy targets, review and performance. Paragraph seven includes the requirements to support the system with competent personnel, resources, awareness, communication and documented information. Paragraph eight discusses operational planning and control, design and procurement requirements. Paragraph nine sets the provisions for the system's performance evaluation and paragraph ten the requirements for process improvement and continuous improvement of the energy management system.[7, 8]

How does ISO 50001 work?

- It identifies **what** requirements the organisation must meet.
- It does **not** identify **how** to meet the requirements.
- Every Energy Management System is unique.
- It is quite general; hence it is flexible.
- Using an EnManSys benefits the organisation.
- It is a framework to manage energy.
- Leads to cost and energy reduction.
- It can create a benchmarking.
- It offers organisational engagement.
- It can support regulatory compliance.
- It can improve organisations

*Figure 9.5* How does ISO 50001 work?

***ISO 45001 Occupational, Safety and Health Management Systems***

This standard defines the requirements for an Occupational Health and Safety (OH&S) management system and guides its use. Organisations can conform to the provisions of this standard in a manner appropriate to their operations, size and nature and commensurate with their

OH&S risks. This standard provides a management tool to achieve the intended purposes of the OH&S management system, improve performance, provide safe workplaces, reduce the risk of occupational injuries, illnesses and fatalities, achieve continual improvement and fulfil legal or other requirements. Aviation facilities, such as maintenance hangars or manufacturing facilities, can benefit from implementing such a management system. They can protect employees through a systemic approach, abiding simultaneously by regulations for the same purpose. The structure of ISO 45001 is the same as all other ISO standards, however, differentiating in scope. ISO 45001 can help an aviation facility to meet its sustainability objectives and goals by preventing and mitigating the risks that will compromise employees' health, productivity and contribution to the overall organisational performance. Employees' absence due to health issues will disrupt the system's operations. At the same time, if occupational and health risks are not adequately mitigated, it could lead to fines or even lawsuits, costing not only money but also affecting reputation and raising concerns about the company's work ethic. ISO 45001 includes ten paragraphs as well. Paragraph four discusses the requirements for the organisation's context. Paragraph five describes the criteria for leadership and worker participation, commitment, occupational, health and safety policy and consultation and participation of workers. Paragraph six sets the requirements for planning, OH&S risks and objectives. Paragraph seven includes the requirements to support the system with competent personnel, resources, awareness, communication and documented information – same requirements as all the other ISO standards but with the OH&S objectives. Paragraph eight discusses operational planning and control, design and procurement requirements. Paragraph nine sets the provisions for the system's performance evaluation and paragraph ten the requirements for process improvement and continuous improvement of the OH&S management system, with incident nonconformity and corrective actions. It is essential to note that all the requirements for the standards explained in this section require to follow in detail the respective documentations.[9]

*Figure 9.6* Recycle bin

**Waste Management**

Waste management systems and practices are essential to be present in all aviation facilities, starting with airports. There are many waste management practices in airports that can support economic and environmental sustainability. When an airport has effective waste management systems and methods, there are positive effects on airport facilities, customers and the local society. Of course, we should remember that a circular economy needs effective and successful practices when discussing waste management.

*Aviation waste* is any unwanted or unused material, product, by-product or substance used, produced or placed at an airport site, terminal or maintenance hangar. The type of waste can vary; it might come from hangars within the airport. In that case, the waste can be hazardous or toxic materials, substances or their containers or from travellers' waste in the terminals. Therefore, the different types of aviation waste require other waste management practices. *An aviation waste management system* includes specific processes and procedures to handle the different kinds of aviation wastes from various facilities' different activities of multiple facilities: an MRO site, an airport terminal, an airport hangar or other aviation-related facilities.

Let us try and provide different aviation waste categories to expand the analysis of aviation waste. If we have specific types of waste with certain characteristics defined, then it is clearer what management practices must be in place. All wastes require different approaches to handling, collecting, disposing or recycling where possible.

- ***Municipal Solid Waste (MSW):*** This waste is the most common daily and disposed of items. They might be steel and aluminium cans, glass and plastic bottles, plastic and paper bags or paper and cardboard products. These wastes are primarily found in airport terminals, flight waste and airline offices. As of ICAO (2018), airports have four main types of MSW.

  i. Terminal waste: waste from travellers and administrative offices in the airport.
  ii. Tenant waste: waste from terminal shops and retail services.
  iii. Airline waste: waste from airplanes and airline offices.
  iv. Cargo waste: waste from cargo operations.

- ***Construction and Demolition Debris (CDD):*** This type of waste, as the word says, is created from projects like excavation, construction, reconstruction and renovations within the airport.
- ***Deplaned or Cabin Waste:*** It is a more specific municipal solid waste that comes from passenger aircraft. This type of waste includes passenger waste, galley waste and waste from the catering process.
- ***International or Quarantined Waste:*** This waste is created from international flights. It is often called quarantine waste since it can include materials and substances from countries with different policies. This waste can contain risks for diseases or other public contaminants called quarantine waste. A possible practice for quarantined waste is to be incinerated or appropriately packaged and sent for disposal from the airport.
- ***Hazardous and Industrial Waste:*** This waste can be chemicals, oils and even aircraft fuel. It can also include solvents, cleaning fluids and substances, de/anti-icing fluids, paint, containers, materials used in aircraft hangars and everything that remains for disposal after maintenance. Disposal of this type of waste is usually based on state or national regulations or unique treatments for storage and/or disposal or recycling.
- ***Lavatory Waste:*** This type of waste comes from aircraft and airport lavatories. They may contain chemicals or human pathogens that can harm public and environmental health and hygiene. That is why they must be carefully disposed of and collected, especially when removed from airplanes' lavatory systems.[10]

***Waste Management Principles***

> 'Airports should promote the culture of avoiding solid waste generation and where possible, extracting value from remaining waste and sending as much as possible or even zero waste to landfills'.[11]

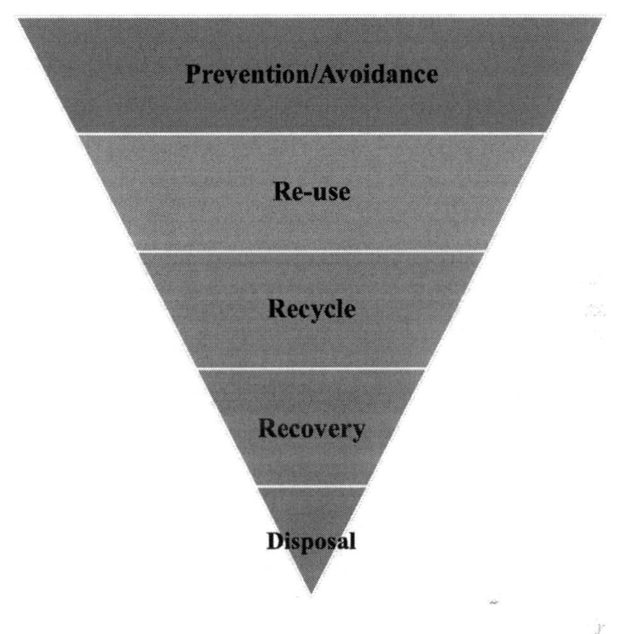


*Figure 9.7* Waste hierarchy

- ***Waste Avoidance:*** This stage refers to the practices that prevent a substance or product from becoming waste.
- ***Waste Reuse:*** Airports can implement waste minimisation activities, reuse materials and reduce the demand for more materials. Waste reduction and reuse can contribute to airports' environmental and economic sustainability; thus, airport operators should have structured programs for waste management.
- ***Waste Recycling:*** The highest share of MSW in airports is 75% paper. Recycling is the most appropriate method for reducing residual waste, reusing where necessary, and reproducing energy.
- ***MSW Recycling:*** If a recycling program is in place for MSW, the benefits are both operational and economic. Staff training is essential in a waste management system, including a recycling program. Specific recorded procedures on how and where to collect the different types of waste are necessary.
- ***CDD Recycling:*** When construction projects are in airports, the waste is construction and demolition debris, as mentioned earlier. We must consider the materials and their recycling in this type of waste. For example, concrete blocks and wooden piping must be appropriately collected and recycled. Airport authorities must consider how to handle this type of debris since the benefits can be economic, environmental, operational and social.
- ***Waste to Energy:*** Waste to Energy (WtE) is a process where non-recyclable waste is converted to a valuable form of energy, heat or electricity or another fuel type. The process is called incineration, when waste is burnt and internal combustion occurs. Anaerobic digestion, gasification and landfill gas recovery convert waste to energy.
- ***Waste Disposal:*** Ultimately, some waste in airports is disposed. Even though there might be waste management systems and recycling programs in place, there might be waste that cannot be reused or recycled.[11–14]

- **Economic Benefits:** Reduced costs for material transportation, disposal fees, fuel consumption, prevent from buying new materials.

- **Environmental Benefits:** Less waste is sent to landfill, reduces the extraction of new resources and materials, reduces fuel consumption, and emissions from transportation of new materials.

- **Operational Benefits:** This process might take place faster, and within the airport, without disturbing the airport operations with delayed accomplishment of tasks due to involvement of external contractors.

- **Social Benefits:** Recycling can benefit local societies with more employment positions and support of local economies. No need for transportation, thus no contribution to traffic.

*Figure 9.8* CDD recycling benefits[11]

**Sustainable Approved Training Organisations**

Going forward with aviation facilities, we must remember that pilot schools are also part of aviation facilities. Pilot schools or Approved Training Organisations (ATO) can and should contribute to a sustainable aviation future. The aviation industry has to follow the net zero targets and all its sectors through their operations. They need to identify ways to comply with the new guidance. ATOs are the least affected organisations or present actions towards flying to net zero.

However, it is essential to mention that when discussing sustainability, we must consider all aspects: environment, society and economy. Starting with the first element, ATOs can significantly contribute to sustainability through the social element. Young pilots are part of the social element, trained in ATOs; therefore, they need to learn from the early times of their studies that they are a significant part of sustainability in the industry.

The pilot students are trained in simulators, having as a central principle safety, and then they get their licenses per aircraft type. However, when discussing environmental sustainability, it is vital to say that international flight operations must follow emission schemes, as we have seen in Chapter 2. The most famous is the ICAO CORSIA since it is global, with EU-ETS following the same principles for the EU region. These schemes are based on recording detailed and accurate information as far as it concerns fuel consumption. Flight times, block-on, block-off times and fuel quantities are some data used in emissions calculations. The pilots are the people who enter this information, sometimes even manually, in very few cases. This information is entered automatically and transmitted to most modern aircraft operators. In their daily routine, professional pilots enter data in the journey log. This daily routine is the starting point of the monitoring, review and verification (MRV) process. However, the verification process does not rely only on the transmitted data. Hence, the information is always crosschecked with the journey logs. Consequently, the more accurate this information is, the more correct the emission calculations are.

It is essential to demonstrate to new pilots from the early times of their careers, even from their training, that they are a significant part of the MRV reporting. All the verified reports are submitted to the state aviation authority where the operator belongs, which is how total aviation emissions are calculated annually. Therefore, the social element contributes to environmental sustainability and sectoral effort to meet emission targets. Another thing that is also part of environmental sustainability to which pilots are contributing is flying practices that are more fuel efficient. Inevitably, this is something that has to do with the pilots' skills. Fuel-efficient flying practices and techniques support environmental sustainability but also support less fuel consumption, which translates to economic savings for the operator. Under that aspect, the operator enhances economic and environmental sustainability without significant changes. Pilots' culture enhances this approach and must begin early in their training school.

Another element where ATOs contribute to sustainability during training is the new simulation technologies used. Synthetic flight training combines real-time simulation and practical flight training. Therefore, even though it is not an approach to meet environmental sustainability, flight hours and fuel consumption are reduced. Of course, it is part of the flight training since the regulation mandates it. However, it contributes to environmental sustainability since we have reduced fuel consumption and flight emissions. Apart from the pilot training, an ATO must operate under sustainability principles as a facility. Operational sustainability in ATO facilities requires using energy and water sensibly, having recycling practices, promoting electronic material with less printed paperwork and contributing to a reduced carbon footprint.

At the same time, facilities can be changed by implementing energy and environmental management systems and having proper waste management approaches through standardised techniques, like ISOs. Environmental and energy systemic pathways will support both environmental and economic benefits for the ATO. Additionally, when we talk about ATOs and training, there are many people, from instructors, visitors, students and whoever is involved with ATOs operations. This kind of initiative can influence the social element.

ATOs are organisations that, at the moment, do not have specific guidance in terms of sustainability. However, their contribution can be immense. As mentioned earlier, these organisations are training new aviation professional pilots, who have a core contribution to emissions, fuel efficiency and consumption. Changing the culture is necessary to meet the pathways to net zero and create professionals with a sustainability mindset.

## Conclusion

Aviation facilities have multiple operations and processes to meet the sector's requirements. Buildings, maintenance facilities, airline offices and airports have various environmental effects, from energy consumption to waste. This chapter explored the role of certification buildings, ISO standards and waste management practices. We discussed many advantages of LEED-certified facilities for an airport, airline office or a maintenance hangar. There are multiple benefits for the environment, the community and the economic state of the building. The great benefits of a LEED certification can enhance internal working conditions, benefiting employees, management, travellers and guests, greatly supporting social sustainability. Energy efficiency and reduced energy consumption reduce operational and energy costs, which inherently support economic sustainability. BREAM certification and ISO standards can help aviation facilities to reduce their carbon footprint and environmental effects and, consequently, have significant expenses from mindless energy consumption. ISO 14001, 50001 and ISO 45001 are tools supporting a sustainable working environment. They are not mandatory; however, if implemented, they must be in place properly, meeting all the requirements. Any aviation facility to be sustainable must use all the appropriate and available techniques and tools. A sustainable organisation should have environmentally concerned operations and offer proper conditions for employees, customers and passengers. The holistic approach of sustainability requires having aviation facilities that will maintain their operational standards but respect, at the same time, all three pillars, the social element, the environment and the economy of the system, inside and outside.

**Key Points to Remember**

- Leadership in Energy and Environmental Design (LEED) is a globally recognised symbol of leadership and sustainability achievement in efficient, healthy, carbon and cost-saving green buildings.
- LEED is beneficial because they address the triple bottom line. They do this by looking at the system holistically and factors in all critical elements (water, health, energy) to produce the best buildings possible.
- LEED buildings use 23% less energy and 28% less water which cuts down on fossil fuels and saves costs. Having a LEED-certified building also influences those around it to help support climate by creating composting and reducing landfill opportunities with 9% less landfill waste and 50% carbon footprint reduction.
- A management system is a series of defined, organisation-wide processes that support for effective decision-making–related actions.
- The ISO standards are not mandatory for companies to get certified.
- Process is a set of interrelated or interacting activities which transforms inputs into outputs.
- Inputs to a process are generally outputs of other processes. Processes in an organisation are generally planned and carried out under controlled conditions to add value.
- Procedure is the specified way to carry out an activity or a process. The procedures can be documented or not.
- The ISO 14001 scheme contributes to sustainability in an organisation by creating an effective environmental management system that benefits the community and assists in reducing nonconformance to environmental regulation.
- According to the International Organization for Standardization, the ISO 14001 model provides a competitive edge through financial efficiencies, increased stakeholder confidence, strategic business aims, encouragement for environmental improvement of suppliers, improvement of community reputation and employee engagement.
- In April 2007, a UNIDO stakeholders meeting decided to ask ISO to develop an international energy management standard. ISO had identified energy management as one of its top five areas for the development of international standards and, in 2008, created a project committee, ISO/PC 242 'energy management', to carry out the work.
- ISO 50001 2018 is the latest standard for implementing a robust energy management system (EnMS).
- Creating an ISO 50001 energy management system ensures either employees, visitors or passengers that there is a commitment to continually improving energy status.
- ISO 45001 standard defines the requirements for an Occupational Health and Safety (OH&S) management system and gives guidance for its use. Organisations are provided flexibility in how to conform to the requirements of this standard in a manner appropriate to each organisation and commensurate with its OH&S risks.
- Waste management systems must be present in all aviation operations. Starting with airports, waste management practices must be present and effective. There are many practices for waste management in airports that can support both economic and environmental sustainability.

*Table 9.1* Acronym rundown

| | |
|---|---|
| LEED | Leadership in Energy and Environmental Design |
| ISO | International Standard Organization |
| BREEAM | BRE'S environmental assessment method |
| QMS | quality management system |
| SMS | safety management system |
| EMS | environmental management system |
| EnMS | energy management system |
| MRO | Maintenance, Repair and Overhaul |
| MSW | Municipal Solid Waste |
| CDD | Construction and Demolition Debris |
| WtE | Waste to Energy |
| UNIDO | United Nations Industrial Development Organization |
| OH&S | Occupational Health and Safety |
| ATO | Approved Training Organisation |

## Chapter Review Questions

*Figure 9.9* Airplane no. 9 pointing right

9.1 What is the LEED certification, and how can it support a sustainable aviation working environment?
9.2 Does LEED certification in aviation buildings contribute to meet the SDGs, and which of them?
9.3 For this question, select a type of an aviation entity (i.e. airline offices, airport terminal, MRO, etc.). Then identify the factors that support the creation of a sustainable working environment. Are ISOs or buildings certifications included? Which ones and why?
9.4 Explain the purpose of getting certified under an ISO 14001. Provide a list with the benefits of an ISO 14001 to an aviation organisation of your choice.
9.5 How does an ISO 14001 support sustainability in aviation and aerospace industry?
9.6 Which of the SDGs are met by an ISO 14001 certification?
9.7 Explain the purpose of getting certified under an ISO 50001.
9.8 What are the benefits of an ISO 50001 to an airline's office premises?
9.9 How does an ISO 50001 support sustainable aviation operations? Provide an analysis covering all three pillars of sustainability.
9.10 Which of the SDGs are met by an ISO 50001 certification?
9.11 Explain an ISO 45001 briefly and if it is necessary for an MRO to get certified under this standard. Identify some potential benefits.
9.12 After answering question 9.11, provide an analysis about the role of ISO 45001 in creating sustainable aviation workplaces.
9.13 Which of the SDGs could be met by an ISO 45001 certification?
9.14 Do you believe that the sick building syndrome is a factor that can affect safety in an aviation working environment?

## Case Study[15, 16]

Heathrow Terminal 2B has been developed using the most modern, sustainable and best practices following the industry's leading innovations to deliver their work sustainably. The constructor, the company responsible for the terminal's construction aiming to achieve high quality and sustainability objectives, worked closely with the supply chain of the terminal and the Heathrow Airport Company. They created a sustainability benchmark, and terminal two of Heathrow became the first U.K. airport with a BREEAM certificate. At the same time, the construction company implemented a range of sustainability initiatives to meet and achieve the BREEAM rating of 'very good' for sustainable construction, design and innovation. Initially, the construction company developed a sustainability action plan, defining the project cycle from the design, procurement and construction phases. For each one of the phases, there were specific approaches that were set to meet sustainability principles. During the design phase, the purpose was to provide optimised design solutions with efficient systems incorporated into the final building design. For example, one innovative design was the fire sprinkler system, which saved 300,000 litres of potable water per year since it was recycling wastewater. In a big project like developing a new airport terminal, the export of material to landfill was expected to be significant. The construction company was committed to reducing, reusing and recycling materials. One of the site's most innovative waste reduction approaches included using over 44,450,000 cubic meters of London clay to remediate a local nonoperational landfill site. On top of that, a significant amount of on-site gravel and concrete were appropriately processed and reused below aircraft stands and the airport taxiways.

At Terminal 2, Lean construction and integrating modularised elements supported minimising work in height, reducing waste, improving day-to-day performance and increasing on-site reliability. Other benefits included:

- Coordinated planning for daily activities.
- Reducing the repetition of work.
- Standing time.
- Minimising changes that would compromise the scheduled operations.

The Terminal 2 project followed approaches for improved quality, health and safety performance. Additionally, the construction company supported the development and establishment of the Heathrow Construction Academy, an initiative to offer training and employment opportunities to people from five nearby boroughs – an excellent initiative for sustainable employment, supporting the local community and the terminal. After reading this case, expand your research and answer the following case study questions.

#### *Case Study Questions*

9.15 What are the most important waste management practices that took place in the terminal's construction?

9.16 Do you find these practices sufficient and why?

9.17 How do the waste management techniques applied potentially to the airport's sustainability?

9.18 The terminal is certified under BREEAM. Do you believe it is sufficient? What are the benefits for the airport under this certification?

9.19 Could you provide some examples of how the local society is benefited from the terminal's construction?

9.20 If the airport authorities considered a CDD recycling method, provide some examples of how CDD could be reused in the airport while in operation.
9.21 Assuming that the new airport structure hinders conditions that are similar to a sick building, how may these conditions affect the airport operations holistically?

## References


[1] [Taipei Taoyuan International Airport]. (2017). https://upload.wikimedia.org/wikipedia/commons/1/1a/Taipei_Taoyuan_International_Airport%2C_Floor.jpg

[2] U.S. Green Building Council. (n.d.). *LEED rating system*. LEED Rating System. Retrieved September 26, 2022, from www.usgbc.org/leed

[3] [LEED Certified Gold]. (2013). https://upload.wikimedia.org/wikipedia/commons/d/db/LEED_Certified_Gold.jpg

[4] International Organization for Standardization. (2015). *ISO 9001:2015 quality management systems – requirements*. International Organization for Standardization

[5] International Organization for Standardization. (2015). *ISO 14001:2015 environmental management systems – requirements with guidance for use*. ISO. Retrieved September 26, 2022, from www.iso.org/standard/60857.html

[6] National Quality Assurance Ltd. Global Certification Body. (2019, February 8). *Guide to ISO 50001*. www.nqa.com/en-us/resources/blog/february-2019/guide-to-iso-50001

[7] Lazarte, M. (2016, November 7). *Does ISO 50001 still live up to its promise?* International Organization for Standardization. Retrieved September 26, 2022, from www.iso.org/2016/11/Ref2135.html

[8] International Organization for Standardization. (2018). *ISO 50001:2018 energy management systems – requirements with guidance for use*. ISO. www.iso.org/standard/69426.html

[9] ISO. (2018). *ISO 45001 requirements for occupational safety and health* (No. 2018–03). ISO

[10] International Civil Aviation Organization. (n.d.). *Waste management at airports | eco airport toolkit*. ICAO| Environment. Retrieved September 26, 2022, from www.icao.int/environmental-protection/documents/waste_management_at_airports_booklet.pdf

[11] International Civil Aviation Organization. (n.d.). *Water management at airports | eco airport toolkit*. ICAO| Environment. Retrieved September 26, 2022, from www.icao.int/environmentalprotection/Documents/Water%20management%20at%20airports.pdf

[12] Federal Aviation Administration. (2022, August 22). *Airport recycling, reuse, and waste reduction*. Retrieved September 26, 2022, from www.faa.gov/airports/environmental/airport_recycling

[13] International Air Transport Association. (n.d.). *Managing cabin waste*. IATA Cabin Waste. Retrieved September 26, 2022, from www.iata.org/en/programs/environment/cabin-waste/

[14] European Aviation Safety Agency. (n.d.). Waste framework directive. *Environment*. Retrieved September 26, 2022, from https://environment.ec.europa.eu/topics/waste-and-recycling/waste-framework-directive_en

[15] Specification Online. (2016, March 22). *The first BREEAM rated airport project*. Retrieved September 26, 2022, from https://specificationonline.co.uk/sections/hotel-sport-and-leisure/articles/2016-03-22/balfour-beatty/the-first-breeam-rated-airport-project

[16] *BRE Group: News*. (n.d.). Retrieved September 26, 2022, from www.bre.co.uk/news/Airbus-lands-a-BREEAM-Excellent-843.html

# 10 Sustainability and Resilience in Aviation and Aerospace

## Chapter Outcomes

At the end of this chapter, you will be able to do the following:

- Learn about future technologies for a sustainable aviation.
- Explain how hydrogen fuel cells work for aircraft.
- Discuss about supersonic technologies.
- Explain new business methods to support aviation sustainability.

## Introduction

As discussed in this book and its previous chapters, sustainability is a relatively new topic for aviation and aerospace operations. The future looks very promising but also unknown. Technology is the number one supporter of sustainable operations in all sectors. Aviation, an industry that relies heavily on technology, is expected to shift significantly towards sustainability. Many new technologies, infrastructures and processes have already entered the aviation sector with significant advancements to be on the verge of changing the current scenery. Hydrogen combustion, new aircraft and aerodynamic structures for reduced fuel consumption and noise levels and supersonic technology are future technological innovations that we expect to see soon as part of a sustainable aviation sector. Nevertheless, we should remember that sustainability starts from a system's culture and openness to change. Innovation and culture go hand in hand for any sustainable system. Thus, sustainability is expected to appear from both these two elements.

## The Future of Aviation

Media lately are full of sustainable aviation advancements. New technologies, designs and some futuristic aircraft are making a solid presence, showing where the industry is heading. There are many projects with numerous technological designs and innovations. Nobody can forget the Solar Impulse 2 project. In 2015, Solar Impulse started the attempt of the first round-the-world solar flight from Abu Dhabi to Hawaii, already achieving the longest solo solar flight ever achieved in aviation history. In 2016, two pilots and founders completed the first circumnavigation of the globe with no fuel. Without a doubt, it was an outstanding achievement. With their solar aircraft, a flying laboratory full of clean technologies, the two pilots flew 40,000 km to promote renewable energies and energy efficiency on the ground for a better quality of life. Now the commercial production and use of solar aircraft depend on many factors, the primary being always safe and reliable transportation. The big question that arises is, 'Are we there yet?' And if not yet, then when? The efforts are numerous and develop rapidly.

DOI: 10.4324/9781003251231-10

Another notable example towards this new era is the Airbus E-Fan X – electric aircraft. It was a hybrid electric aircraft demonstrator being developed by a partnership of Airbus, Rolls-Royce PLC and Siemens. It was announced in November 2017 and followed previous electric flight demonstrators towards sustainable transport. Even though this one was cancelled in April 2020, it is still an example that shows that aerospace aims to follow sustainability, support new designs and make significant progress. Naturally, these are only two examples of the many new technologies and designs that are changing aviation as we know it. However, the change is from more than just the aircraft design and structure perspective. It also includes new platforms and services that support aviation services, primarily on the flying part. Research and development aim to progress with collaborative schemes to make electric vehicles available for transportation but always under safety.

One more noteworthy case of a futuristic design is an innovative demonstrator called Maveric. Maveric has been developed by Airbus, standing for Model Aircraft for Validation and Experimentation over Robust Innovative Controls. It is 2 meters long and 3.2 meters wide, with a surface area of about 2.25 square meters. Maveric's design can reduce fuel consumption by up to 20% compared to current single-aisle aircraft. Although there is no specific timeline for entering into service, this technological demonstrator could be instrumental in changing commercial aviation architectures for an environmentally sustainable future for aviation.

The last example, VoltAero Cassio, is a hybrid electric demonstrator. The Cassio 1 is based on a Cessna 337 model. Its original forward-mounted internal combustion engine has been removed and replaced by a smooth nose, enabling it to go airborne with the power of two Safran electric motors installed on the wings and the aft-facing power module in the pusher position.

Naturally, these examples are only a few currently under research and development. Models are developed, tested, approved for continuation or rejected from the early stages of trial, all under consideration for changing the aviation world. However, some are already applied, like the Airbus A320/321neo.

*Figure 10.1* Sharklets[1]

The A320neo, where neo stands for new engine option, offers 15% fuel burn savings compared to the current single-aisle aircraft operations and is said to increase to 20% in the future spurred by further developments to the new engine options, sharklets and carbon optimisations. Nevertheless, someone might ask: What are the sharklets? Sharklets are components added at the wing tip. Adding the sharklets at the wing is a way to increase their dynamic span and reduce the tip vortex created by the imbalance of pressure in the upper and lower surface of the wings, delivering up to 4% in fuel burn savings.

*Figure 10.2* Airbus 320neo[2]

Zephyr is a HAPS, a high-altitude pseudo satellite. It is the world's leading solar electric, stratospheric uncrewed aerial vehicle. It is a technology appropriate for humanitarian and environmental missions worldwide. Zephyr harnesses the sun's rays, running exclusively on solar power above the weather and conventional air traffic.

**Supersonic Flights**

Supersonic flights are part of a technology that belongs to the past; however, there are various research efforts to bring them back into action without the harmful effects on the environment and the high cost of tickets and operations. Under these circumstances, before exploring and seeing how viable it will be to return supersonic flights to operation, let us see what the *supersonic* flights are.

Supersonic flights travel through the air at a speed greater than the sound speed at a specific location. This means that the speed of sound, which in aviation language refers to as Mach 1, varies with atmospheric pressure and temperature. For example, when the air temperature is 15°C (59°F) and sea-level pressure, the sound travels at 1,225 km (760 miles) per hour. Hypersonic flight is used when the speed is about five times the speed of sound (Mach 5). Any flying vehicle or object that travels through the Earth's atmosphere at supersonic speeds generates a sonic boom. A *sonic boom* is a shock wave that sounds like a deafening explosion when heard from someone on the ground.

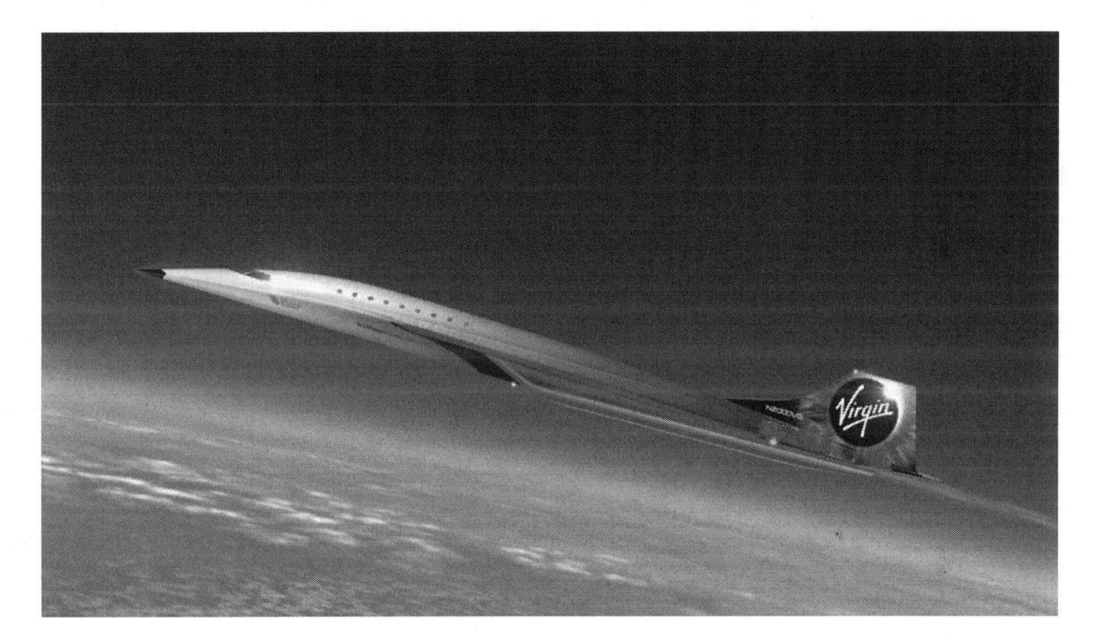


*Figure 10.3* Virgin Galactic[3]

The supersonic aircraft created a big sonic boom that often disturbed the surrounding environment, which was a big drawback. In addition, fuel consumption is five to seven times higher than conventional fuel. So why are we looking back to a technology that was not so sustainable? New projects about bringing back supersonic flights are back in place from big players in the aerospace industry. So what has changed from thirty or forty years back? The significant change in the market lies in the newly introduced SAF and its prospective usage in supersonic flights, along with more advanced and efficient technologies and design.

A Bell XS-1 rocket-powered research plane, piloted by Major Charles E. Yeager of the U.S. Air Force on October 14, 1947, was the first supersonic speed aircraft. The XS-1 broke the (local) sound barrier at 1,066 km (662 miles) per hour and attained a top speed of 1,126 km (700 miles) per hour, or Mach 1.06, after it was dropped from the belly of a Boeing B-29 mother ship. After that, many military aircraft capable of the supersonic flight was built, though their speed was generally limited to Mach 2.5 because of problems caused by frictional heating of the plane's skin.

The first supersonic passenger-carrying commercial airplane (or supersonic transport, SST), the Concorde, was built jointly by Aérospatiale and British Aerospace. The Concorde was the first cooperation between the two countries to design an aircraft. On November 29, 1962, Britain and France signed a treaty to share the costs and risks of producing an SST. British Aerospace and the French firm Aérospatiale were responsible for the airframe. At the same time, Britain's Rolls-Royce and France's SNECMA (Société Nationale d'Étude et de Construction de Moteurs d'Aviation) developed the jet engines. The result was a technological masterpiece that made its first flight on March 2, 1969. The Concorde's maximum cruising speed of Mach 2.04 (more than twice the speed of sound) allowed the trip duration from London to New York to reduce by about three hours. On January 21, 1976, the Anglo-French Concorde inaugurated the world's first scheduled supersonic passenger service. British Airways initially flew the aircraft from London to Bahrain, and Air France flew it from Paris to Rio de Janeiro. Both airlines added regular service to Washington, D.C., in May 1976 and New York City in November 1977. Other routes were added temporarily or seasonally, and the Concorde was flown on chartered flights to destinations worldwide. Concorde's maximum cruise speed was 2,179 km (1,354 miles) per hour, or Mach 2.04. The first concord flew on March 2, 1969, made its first flight crossing the Atlantic on September 26, 1973, and entered into regular service in 1976. The aircraft proved a technical miracle but, eventually, an economic disaster. Never financially profitable, it was retired in 2003.[4] Since the 1960s, England and France have flown Concorde supersonic transports (SSTs) as passenger airlines since 1960s. Supersonic passenger transport ended on July 25, 2000. The Air France Concorde took off from Airport Charles de Gaulle, and while on the runway, it hit some debris. One of the tires burst, and more debris fractured a wing fuel tank. Almost instantly, the plane erupted in flames. The Concorde rose briefly and then it crashed into a hotel. Fatalities accounted for 109 people on the plane and four on the ground. At the time of the accident, Concordes were already being retired, which permanently grounded the remaining supersonic fleet.[5]

Despite the great convenience of supersonic flights in air transportation, the implications remained high and very serious. Noise levels throughout all stages of flight were not only harmful to the natural environment while taking off and landing but also disturbing the residents and the nearby natural environment. The cost of a Concorde flight was also incredibly high. However, the reduced time of long-haul trips superseded all other adverse outcomes. After the cease of operations of all Concorde flights in 2000, there are new efforts to bring back supersonic flights. However, since the aviation industry is on a pathway to sustainability, all previous impactions of supersonic flights – environmental, social and safety related – must be highly evaluated and minimised. If we want a re-entrance of supersonic aircraft in the air transportation system, it must be a technological advancement that will advocate a sustainable aviation sector.

*Figure 10.4* The Concorde[6]

## Hydrogen Cells

Hydrogen is one of the most abundant chemical elements, and one of its most valuable uses is in hydrogen fuel cells. A fuel cell combines hydrogen and oxygen for electricity generation and water's chemical formation. Hydrogen's use has become increasingly important, given the necessity to store and use energy without greenhouse gas emissions. Until now, fuel cells have been used in almost all mobility means: trains, cars, buses, ships, submarines, forklifts, even bicycles, etc., except in civil aviation. Fuel cells could replace secondary power systems, such as auxiliary power units (APUs) and main engine-driven generators (EDGs).

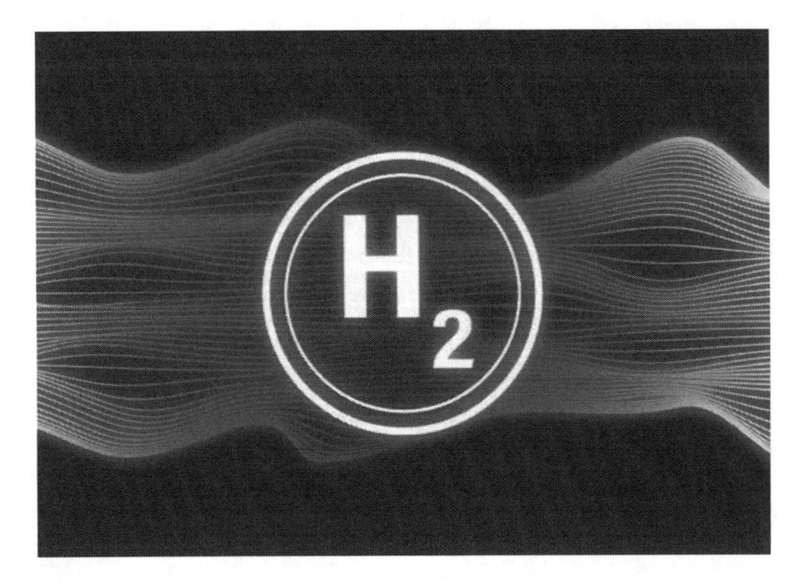


*Figure 10.5* Hydrogen molecule

A few successful cases of fuel cell applications in commercial airplanes and general aviation have already been performed. The hydrogen fuel cell is a technology to reach the net zero goals for climate change and energy security in several commercial sectors, including transport, industry and power generation/distribution. In addition, connecting different hydrogen-using sectors with transmission and distribution networks will increase the operational flexibility of the future low-carbon energy economy. Hydrogen is becoming an alternative to hydrocarbons as a high-quality energy source for aircraft applications. Hydrogen is plentiful, non-polluting, renewable and efficient to use. Hydrogen (H) is non-toxic, non-polluting, colourless and odourless. It is the simplest and smallest atom; two atoms will make the gas molecule (H2). Because of this, it has the lowest density of any gas. For example, 1 m3 of air at sea level weighs 1.3 kg, whereas 1 m3 of hydrogen weighs 14 times less at 0.09 kg. This means that hydrogen rises and dissipates naturally – so it diffuses faster than any other gas, with a high risk in case of leakage. When hydrogen reacts with oxygen – an oxidation process – energy is released. The process is as in internal combustion engines or through a controlled electrochemical reaction, as in fuel cells. The hydrogen combustion products are heat and water. In the case of a fuel cell, electrical energy is also produced along with heat and moisture.

Hydrogen has the lowest fuel density but the highest energy content by mass. However, the volume required to store gaseous hydrogen is higher than other liquid fuels; hydrogen has three times the energy content of jet fuel (kerosene) by mass but requires considerably more volume (both gaseous and liquid). Hydrogen does not readily self-ignite because of its high auto-ignition temperature of 585°C. The hydrogen's flammability range varies from 4% to 75% by volume in the air; the opportunity for a fire to occur can be managed, as with any fuel. For instance, hydrogen is stored as a pure substance in its container – unlike jet fuel (kerosene), which is mixed with air, eliminating the combustion risk in the container itself.

Hydrogen combustion produces up to 90% fewer nitrogen oxides than kerosene fuel, eliminating particulate matter formation.[7] From an environmental and energy content perspective, hydrogen has abundant potential. The criteria that define a fuel's quality are high energy density, inexhaustibility, cleanliness, convenience and independence from outside sources.[8] Liquid hydrogen meets these criteria, along with the potential to eliminate emissions from combustion. Another valuable characteristic of hydrogen is that it can replace liquid fuel or be used as a fuel cell for electrical power. Electrical fuel cells could be suitable for short-range aircraft, while hydrogen combustion would be suitable for long-range and higher payloads. Hydrogen fuel cells are already standard devices found in cars, buses and often in aircraft servicing vehicles.[9] Liquid hydrogen has a lower volumetric density than kerosene. It is estimated that to complete a given mission, despite the aircraft requiring a lower fuel mass, the space this fuel would occupy would be around four times larger than that of kerosene.[10] This could be a challenge for aircraft and airframe designers, and it would demand a significant redesign of current airframes.

Water vapour is another greenhouse gas produced by fuel combustion. Radiative forcing – the difference between the energy absorbed through the Earth's atmosphere and the energy reflected into space – has a lower value than CO2. However, it still affects global warming. Hydrogen combustion produces about 2.6 times more water vapour than kerosene combustion. CO2 molecules have a longer lifetime in the atmosphere of up to 100 years. Conversely, water vapour can last from a few days to 1 year. When burning hydrogen, solid particles are absent at the engine's exhaust. The ice crystals formed have nowhere to nucleate. Hence, the number of water crystals formed in the engine's exhaust would decrease. Nevertheless, due to the increased amount of water vapour exhaust, the crystals that do nucleate would have a bigger size. Ultimately, the overall effect would decrease the radiative forcing effect of contrails. This signifies that the radiative forcing from aviation could be 20–30% lower by 2050 and 50–60% by 2100 if LH2 aircraft were introduced.

### Electric Aircraft

The newest and up-and-coming technology for sustainable aviation is electric aircraft. Electric aircraft can reduce emissions compared to aircraft that use and consume traditional fossil fuels. Electric aircraft rely on electric motors powered by batteries or fuel cells, which do not produce emissions during operation. In contrast, conventional aircraft use fossil fuels, which release greenhouse gases and other pollutants into the atmosphere, contributing to global warming and air pollution. In addition, the noise pollution caused by traditional aircraft can also negatively impact the environment and human health. While electric aircraft still require energy to charge their batteries or fuel cells, the sources of this energy can be renewable, such as solar or wind power, which further reduces emissions.

*Figure 10.6* An electric aircraft[11]

However, it is essential to note that the production and disposal of batteries and other components used in electric aircraft can also have environmental impacts, such as the extraction of rare earth metals and the potential for toxic waste. Efforts are being made to mitigate these impacts through improved recycling and disposal processes and more sustainable manufacturing materials.

Overall, while electric aircraft are not a silver bullet solution to the aviation industry's emissions problem, they can significantly reduce emissions and promote sustainable aviation.

**New Tools for Aviation Sustainability**

Besides new aircraft designs, structures, flight operations and sustainable fuels, new options are available to support aviation sustainability. Fuel efficiency platforms, booking of SAF and even a concept of carbon exchange are some options we need to value when we want to bring sustainability to aviation. The SkyBreathe platform, the Book and Claim solution and the Aviation Carbon Exchange are explained next to show where aviation sustainability goes.

The SkyBreathe fuel efficiency platform uses big data algorithms and artificial intelligence, with machine learning to automatically analyse data points from flight recorders. The software identifies the most relevant saving opportunities by combining them with environmental data from actual flight conditions. It provides a series of recommended actions which can reduce total fuel consumption by up to 5%. SAS Ireland, Norwegian, Malaysia Airlines, Transavia and Atlas Air are some of the many large airliners already using it, taking benefits from an accurate understanding of fuel users through all flight phases to implement the most efficient procedures without compromising safety. This platform is only one example from numerous already in place, and it is a practice towards fuel efficiency and reduced environmental effects from flights.

'Book and Claim' is a solution that could enable airlines to purchase SAF without being geographically connected to a supply site and further transfer its sustainability attributes to their corporate partners. From a technical perspective, 'Book and Claim' is a chain of custody model that allows to decouple the consumer for the actual physical product while still allowing the consumer to claim the CO2 reduction their purchase achieves.

SAF is not physically transported and entered a specific aircraft of an airline. Instead, it goes into the fuel system of an airport close to the SAF facility. The volume of SAF produced and entered into the airport's fuel system is tracked and verified, after which corresponding carbon emissions factors are calculated and allocated to the airline that covers the expense.

The 'Book and Claim' system offers a series of benefits:

- **More efficient and sustainable supply chains.** SAF entering the airport's fuel system near the production facility minimises Scope 3 GHG emissions related to SAF.
- **Airline and location independent system.** The 'Book and Claim' system allows corporate buyers to source SAF based on their total aviation footprint in one transaction rather than sourcing through each airline individually. SAF can be sourced for flights with airlines or out of airports without SAF supply.
- **Flexible purchase.** Companies can purchase SAF without technical limitations, such as blending limits.

Although investing in out-of-sector carbon reduction should not be considered the primary means of airlines' meeting long-term emissions goals, offsetting will still be needed to remove residual CO2 emissions once all technological, operational and infrastructure improvements have been implemented. The IATA Aviation Carbon Exchange (ACE) is a centralised marketplace for CORSIA eligible emission units. Airlines and other aviation stakeholders can trade CO2 emission reductions for compliance or voluntary offsetting purposes. ACE offers a secure and easy-to-use trading system with the highest transparency in terms of price and availability of emissions reduction. ACE was launched in 2020 and is currently available for IATA and non-IATA members and accessible to carbon market participants aiming to list emissions reduction under CORSIA.

**Closing the Loop: What Is Aviation Sustainability?**

Closing the loop of the information presented in this book, inevitably, aviation sustainability is a new topic that requires addressing all the elements that form aviation operations. Environmental sustainability is one of the favourite topics the industry currently focuses on. Sustainable Aviation Fuels, new aircraft designs and technologies are the most well-known elements to create an industry that will meet environmental sustainability. However, environmental sustainability is only one aspect since we must look at the aviation system holistically.

As shown in Figure 10.7, a sustainable aviation system must consider all three aspects equally and in balance. Figure 10.7 shows all three sustainability elements for aviation, identifying what an aviation system needs to include for operating under sustainability principles. The first one is about environmental sustainability. Environmental and waste management practices in airline offices, airports, pilot schools and Maintenance, Repair and Overhaul (MRO) facilities are essential to implement approaches for reduced environmental effects. In addition, necessary activities include water conservation and pollution control in airports and maintenance facilities. As mentioned in previous chapters, emission control in flights through schemes like CORSIA and EU- ETS is also important and necessary when we talk about environmental sustainability. Sustainable Aviation Fuels (SAF), airport noise control measures, new aircraft designs and engines for reducing emissions and noise are significant environmental sustainability attributes.

Moving on with social sustainability for aviation is also critical. Social sustainability for aviation means considering the social element, which comprises employees and the people professionally involved with aviation operations, passengers and customers. In aviation social sustainability, we must appreciate the social component by respecting the aviation regulation and providing high safety standards. Additionally, support of aviation operations with proper resources to maintain the viability and functionality of several management systems, such as safety, quality, cybersecurity or any other related systemic functions, are also part of social sustainability. Additionally, social sustainability mandates support the currency in the professional development of pilots, maintenance technicians, practical and theoretical instructors, air traffic controllers and more. Since aviation is a sector that follows and uses technology, personnel must continually be updated and appropriately trained in new technologies. Ethical considerations for employees and passengers are also necessary for a socially sustainable aviation environment. All companies in different areas must aim to employ locals to support the society economically, respecting, at the same time, nearby residents in airport and aircraft operations through reduced noise and disturbance measures.

A crucial element and part of aviation social sustainability is also culture. Aviation sustainability requires shifting the culture to all aviation professionals to create a sustainable aviation industry.

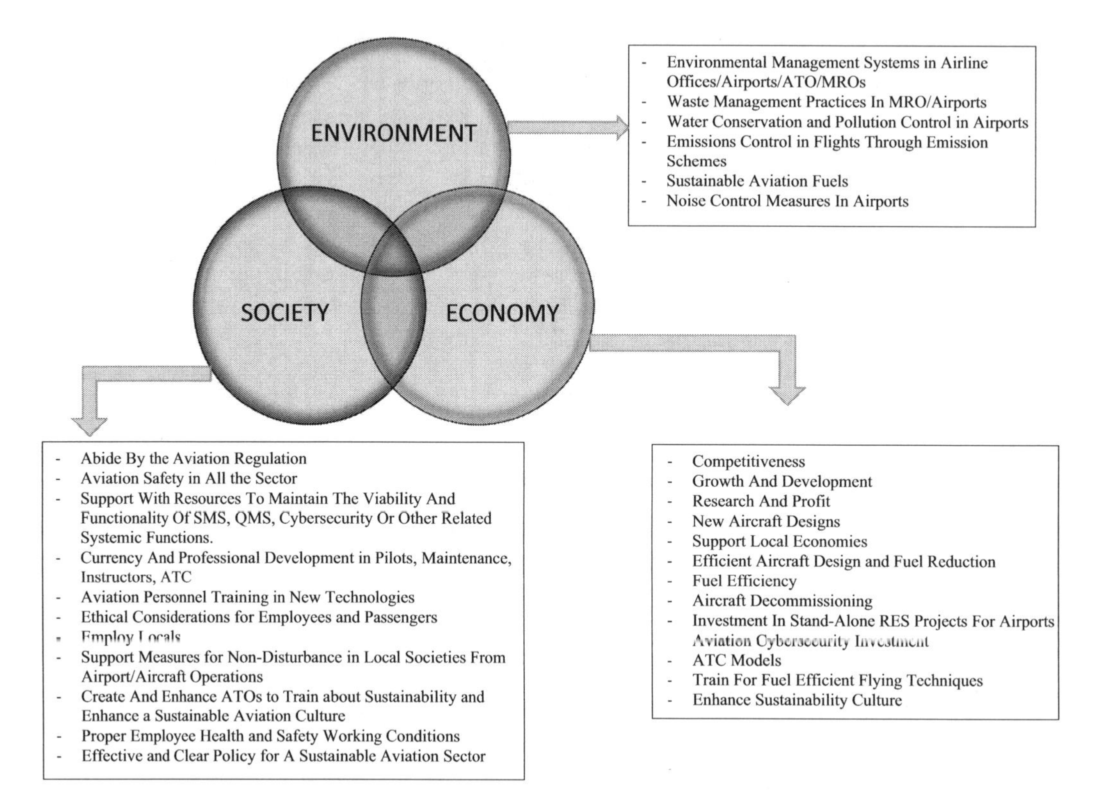


*Figure 10.7* Aviation sustainability model

Pilot schools and aviation maintenance training organisations must start training young professionals about sustainability and their contribution to developing a sustainable aviation culture. Finally, as part of social sustainability, proper employee health and safety working conditions must be valued in all aviation sectors, considering the role of employees' mental health significantly.

Finally, economic sustainability for aviation means aiming for continuous improvement, profit and constant growth. New designs are necessary for safer and faster flights supporting the industry's operational and economic status. Aviation economic sustainability seeks to support local economies. A system must be economically sustainable not only internally but also externally. Efficient aircraft design in new airplanes will reduce fuel and costs. Fuel efficiency is also an element to support economic sustainability in aviation. *Aircraft commissioning* is a new topic that can significantly support aviation's circular economy and economic sustainability. Aircraft commissioning can help reuse or sell materials, equipment or components to other aviation-related facilities and companies and industries where they can use redundant materials from decommissioned aircraft. Additionally, airports can develop renewable energy projects by installing nearby stand-alone photovoltaic systems to cover their electricity or sell the excess electricity produced.

As mentioned earlier, the support of cybersecurity is not only part of social sustainability but also part of economic sustainability. It can prevent and save companies, airports or airlines from cyberattacks that can lead to trust issues or affect their commercial image and operations. Air traffic control models are also necessary to support economic sustainability because they will make employees work more efficiently, avoid flight delays, maintain safety standards and keep passengers satisfied. Finally, fuel-efficient flying techniques are also part of economic, social and environmental sustainability. Fuel-efficient flying techniques mean reduced fuel consumption and reduced emissions based on pilots' skills and training.

To conclude, sustainability is a holistic concept that must be embraced and applied in all aviation operations, accepted by employees and supported by the leadership. Aviation sustainability is a topic that concerns everyone who is part of that industry and its operations. Everyone has a responsibility through their work in a company and organisation. Their contribution is immense to creating a sustainable system with a strong culture supported by the management system.

## Conclusion

Closing the journey we started in this book, in this chapter, we examined various types of methods, techniques and technologies, showing the direction of the aviation sector to a sustainable future. Hydrogen technologies and electric aircraft are in the development process. Many models are already in test flights, significantly contributing to reduced noise, fewer emissions and fuel consumption. Supersonic flights were not considered a sustainable option; however, the new release of supersonic aircraft is considered to re-enter the aviation market with technologies with less noise and fuel consumption. Supporting mechanisms like 'Book and Claim' could enable airlines to purchase SAF without being geographically connected to a supply site and transfer its sustainability attributes to their corporate partners. Aviation Carbon Exchange (ACE) can help airlines and other aviation stakeholders to trade CO2 emission reductions for compliance or voluntary offsetting purposes. As presented in all chapters of this book, sustainability requires a series of operations and actions applicable to all areas of the aviation sector. Environment, society and economy should be considered within the sector's operations and outcomes as part of sustainability's holistic approach. The interrelation of aviation processes, human, the environment and the industry's economic viability requires an adaptation to new technologies, culture and operations that will support a sustainable future for the aviation sector.

## Key Points to Remember

- New technologies, new designs and some very futuristic aircraft are making a string presence, showing where the industry heads to.
- In 2015, Solar Impulse started the attempt of the first round-the-world solar flight from Abu Dhabi to Hawaii, already achieving the longest solo solar flight ever achieved in aviation history.
- Research and development aim to progress daily with collaborating schemes to make electric vehicles available for transportation and safe operations.
- The A320neo, where neo stands for new engine option, offers 15% fuel burn savings compared to the current single-aisle aircraft operations and is said to increase to 20% in the future spurred by further developments to the new engine options, sharklets and carbon optimisations.
- Sharklets are components added at the wing tip. But why do we care about sharklets? Adding the sharklet at the wing is a way to increase their dynamic span and reduce the tip vortex created by the imbalance of pressure in upper and lower surface of the wings, delivering up to 4% in fuel burn savings.
- Supersonic flights are back in the research market. The supersonic aircraft creates a big sonic boom that often disturbed the surrounding environment, and this was a big drawback. In addition, the fuel consumption is five to seven times higher than that of conventional fuel.
- Noise levels throughout all stages of flight were not only harmful to the natural environment while taking off and landing but it was also disturbing to the residents and the nearby natural environment.

- Hydrogen is one of the most abundant chemical elements and one of its most valuable uses is in hydrogen fuel cells. A fuel cell combines hydrogen and oxygen for electricity generation and water's chemical formation.
- Hydrogen (H) is non-toxic, non-polluting, colourless and odourless. It is the simplest and smallest atom; two atoms will make the gas molecule (H2).
- Hydrogen has the lowest fuel density but the highest energy content by mass. However, the volume required to store gaseous hydrogen is higher than other liquid fuels; hydrogen has three times the energy content of jet fuel (kerosene) by mass but requires considerably more volume (both gaseous and liquid).
- The newest and up-and-coming technology for sustainable aviation is electric aircraft. Electric aircraft can reduce emissions compared to aircraft that use and consume traditional fossil fuels. Electric aircraft rely on electric motors powered by batteries or fuel cells, which do not produce emissions during operation.

## Acronyms

*Table 10.1* Acronym rundown

| | |
|---|---|
| NEO | new engine option |
| HAPS | high-altitude pseudo satellite |
| SAF | Sustainable Aviation Fuels |
| SNECMA | Société Nationale d'Étude et de Construction de Moteurs d'Aviation |
| SST | supersonic transport |
| APU | Auxiliary Power Unit |
| EDG | energy-driven generator |
| ACE | Aviation Carbon Exchange |
| RES | Renewable Energy Sources |
| ATC | air traffic control |
| ATO | Approved Training Organizations |
| MRO | Maintenance, Repair and Overhaul |
| SMS | safety management systems |
| QMS | quality management systems |

## Chapter Review Questions

10.1 Identify and explain two representative cases of aircraft that are showing the future of the aviation.

10.2 What is the Solar Impulse, and how can this kind of project contribute to aviation sustainability?

10.3 What is a supersonic flight? What supersonic flights stopped operating?

10.4 What are some changes that need to take place so they can be part of a sustainable aviation sector?

10.5 What are the hydrogen fuel cells, and why do they constitute a sustainable option as an aviation fuel?

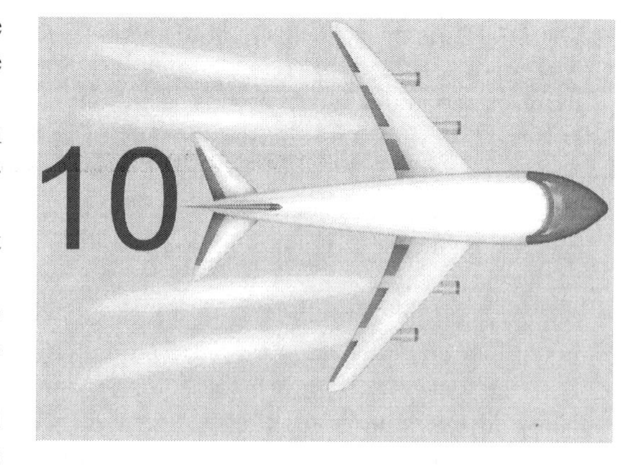


*Figure 10.8* Airplane no. 10 pointing right

10.6 What type of greenhouse gases are emitted from hydrogen combustion?
10.7 Explain if electric aircraft can be considered a sustainable option for aviation.
10.8 What is the SkyBreathe platform, and how can it contribute to aviation sustainability?
10.9 What is 'Book and Claim', and what is the purpose of its use?

## References


[1] Drozdp. (2012, June 11). *Winglet*. Wikipedia Commons. https://commons.wikimedia.org/wiki/File:Winglet_01.JPG

[2] Don VIP. (2014, September 25). *Airbus A320neo landing*. Wikimedia Commons. Retrieved March 21, 2023, from https://commons.wikimedia.org/wiki/File:Airbus_A320neo_landing_04.jpg

[3] Virgin. (2020, June 9). *Virgingalactic*. Wikimedia Commons. Retrieved March 21, 2023, from https://commons.wikimedia.org/wiki/File:Virgingalactic.jpg

[4] Britannica, E. P. S. (2011). *The complete history of aviation: From ballooning to supersonic flight*. Rosen Publishing Group

[5] Garrett, M. (2014). Airplane disasters. In *Encyclopedia of transportation: Social science and policy* (Vol. 1, pp. 102–104). SAGE Publications, Inc. https://dx.doi.org/10.4135/9781483346526.n38

[6] Marmet, E. (1986, May 1). *British airways Concorde G-BOAC 03.jpg*. Wikimedia Commons. Retrieved March 21, 2023, from https://commons.wikimedia.org/wiki/File:British_Airways_Concorde_G-BOAC_03.jpg

[7] SAE International. (2019, November 18). *AIR7765: Considerations for hydrogen fuel cells in airborne applications – SAE international*. SAE Mobilus. www.sae.org/standards/content/air7765/

[8] IATA. (2022). *Fact sheet 7: Liquid hydrogen as a potential low- carbon fuel for aviation*. IATA. www.iata.org/contentassets/d13875e9ed784f75bac90f000760e998/fact_sheet7-hydrogen-fact-sheet_072020.pdf

[9] Bauen, A., Bitossi, N., German, L., Harris, A., & Leow, K. (2020). Sustainable aviation Fuels: Status, challenges and prospects of drop-in liquid fuels, hydrogen and electrification in aviation. *Johnson Matthey Technology Review*, *64*(3), 263. https://doi.org/10.1595/205651320x15816756012040

[10] IATA. (2022). *Aviation carbon exchange*. IATA – ACE- Aviation Carbon Exchange. www.iata.org/en/programs/environment/ace/

[11] File: ROLLS ROYCE PLC ELECTROFLIGHT NXTE G-NXTE (52243151069).jpg – Wikimedia Commons. (2022, July 17). https://commons.wikimedia.org/wiki/File:ROLLS_ROYCE_PLC_ELECTROFLIGHT_NXTE_G-NXTE_(52243151069).jpg

[12] Solar Impulse. (n.d.). *Skybreathe – solar impulse efficient solution*. https://solarimpulse.com/solutions-explorer/skybreathe

[13] Kok, R. (2022, July 14). *SAF book and claim – EBAA – European business aviation association*. EBAA – European Business Aviation Association. www.ebaa.org/press/saf-book-and-claim/

[14] IATA. (n.d.). *Aviation carbon exchange*. www.iata.org/en/programs/environment/ace/

# Index